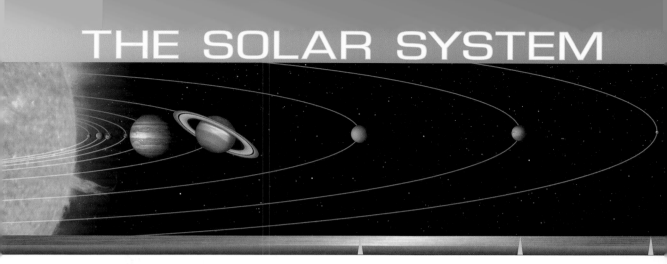

THE SOLAR SYSTEM

URANUS, NEPTUNE, PLUTO,
AND THE OUTER SOLAR SYSTEM

REVISED EDITION

Linda T. Elkins-Tanton

Facts On File
An imprint of Infobase Publishing

In memory of my brother Thomas Turner Elkins,
who, when I was 10 years old, taught me about the Oort cloud,
and together we named our pet mouse Oort.

Uranus, Neptune, Pluto, and the Outer Solar System, Revised Edition

Copyright © 2011, 2006 by Linda T. Elkins-Tanton

Facts On File, Inc.
An imprint of Infobase Publishing
132 West 31st Street
New York NY 10001

Library of Congress Cataloging-in-Publication Data
Elkins-Tanton, Linda T.
    Uranus, Neptune, Pluto, and the outer solar system / Linda T. Elkins-Tanton ; foreword, Maria T. Zuber—
Rev. ed.
        p. cm.— (The solar system)
    Includes bibliographical references and index.
    ISBN 978-0-8160-7701-4
    1. Uranus (Planet) 2. Neptune (Planet) 3. Pluto (Dwarf planet) 4. Solar system I. Title.
    QB681.E45 2011
    523.47—dc22   2010003273

Facts On File books are available at special discounts when purchased in bulk quantities for businesses, associations, institutions, or sales promotions. Please call our Special Sales Department in New York at (212) 967-8800 or (800) 322-8755.

You can find Facts On File on the World Wide Web at http://www.factsonfile.com

Excerpts included herewith have been reprinted by permission of the copyright holders; the author has made every effort to contact copyright holders. The publishers will be glad to rectify, in future editions, any errors or omissions brought to their notice.

Text design by Annie O'Donnell
Composition by Hermitage Publishing Services
Illustrations by Dale Williams
Photo research by Elizabeth H. Oakes
Cover printed by Bang Printing, Brainerd, Minn.
Book printed and bound by Bang Printing, Brainerd, Minn.
Date printed: November 2010
Printed in the United States of America

10 9 8 7 6 5 4 3 2 1

This book is printed on acid-free paper.

Contents

Appendix 2: Light, Wavelength, and Radiation   204

Appendix 3: A List of all Known Moons   214

foreword

While I was growing up, I got my thrills from simple things—one was the beauty of nature. I spent hours looking at mountains, the sky, lakes, et cetera, and always seeing something different. Another pleasure came from figuring out how things work and why things *are* the way they *are*. I remember constantly looking up things from why airplanes fly to why it rains to why there are seasons. Finally was the thrill of discovery. The excitement of finding or learning about something new—like when I found the Andromeda galaxy for the first time in a telescope—was a feeling that could not be beat.

Linda Elkins-Tanton's multivolume set of books about the solar system captures all of these attributes. Far beyond a laundry list of facts about the planets, the Solar System is a set that provides elegant descriptions of natural objects that celebrate their beauty, explains with extraordinary clarity the diverse processes that shaped them, and deftly conveys the thrill of space exploration. Most people, at one time or another, have come across astronomical images and marveled at complex and remarkable features that seemingly defy explanation. But as the philosopher Aristotle recognized, "Nature does nothing uselessly," and each discovery represents an opportunity to expand human understanding of natural worlds. To great effect, these books often read like a detective story, in which the 4.5-billion year history of the solar system is reconstructed by integrating simple concepts of chemistry, physics, geology, meteorology, oceanography, and even biology with computer simulations, laboratory analyses, and the data from the myriad of space missions.

Starting at the beginning, you will learn why it is pretty well understood that the solar system started as a vast, tenuous ball of gas and dust that flattened to a disk with most of the mass—the future Sun—at the center. Much less certain is the transition from a dusty disk to the configuration with the planets, moons, asteroids, and comets that we see today. An ironic contrast is the extraordinary detail in which we understand some phenomena, like how rapidly the planets formed, and how depressingly uncertain we are about others, like how bright the early Sun was.

Once the planets were in place, the story diverges into a multitude of fascinating subplots. The oldest planetary surfaces preserve the record of their violent bombardment history. Once dismissed as improbable events, we now know that the importance of planetary impacts cannot be overstated. One of the largest of these collisions, by a Mars-sized body into the Earth, was probably responsible for the formation of the Earth's Moon, and others may have contributed to extinction of species on Earth. The author masterfully explains in unifying context the many other planetary processes, such as volcanism, faulting, the release of water and other volatile elements from the interiors of the planets to form atmospheres and oceans, and the mixing of gases in the giant planets to drive their dynamic cloud patterns.

Of equal interest is the process of discovery that brought our understanding of the solar system to where it is today. While robotic explorers justifiably make headlines, much of our current knowledge has come from individuals who spent seemingly endless hours in the cold and dark observing the night skies or in labs performing painstakingly careful analyses on miniscule grains from space. Here, these stories of perseverance and skill receive the attention they so richly deserve.

Some of the most enjoyable aspects of these books are the numerous occasions in which simple but confounding questions are explained in such a straightforward manner that you literally feel like you knew it all along. How do you know what is inside a planetary body if you cannot see there? What makes solar system objects spherical as opposed to irregular in shape? What causes the complex, changing patterns at the top of Jupiter's atmosphere? How do we know what Saturn's rings are made of?

When it comes right down to it, all of us are inherently explorers. The urge to understand our place on Earth and the extraordinary worlds beyond is an attribute that makes us uniquely human. The discoveries so lucidly explained in these volumes are perhaps most remarkable in the sense that they represent only the tip of the iceberg of what yet remains to be discovered.

—Maria T. Zuber, Ph.D.
E. A. Griswold Professor of Geophysics
Head of the Department of Earth,
Atmospheric and Planetary Sciences
Massachusetts Institute of Technology
Cambridge, Massachusetts

Preface

On August 24, 2006, the International Astronomical Union (IAU) changed the face of the solar system by dictating that Pluto is no longer a planet. Though this announcement raised a small uproar in the public, it heralded a new era of how scientists perceive the universe. Our understanding of the solar system has changed so fundamentally that the original definition of *planet* requires profound revisions.

While it seems logical to determine the ranking of celestial bodies by size (planets largest, then moons, and finally asteroids), in reality that has little to do with the process. For example, Saturn's moon Titan is larger than the planet Mercury, and Charon, Pluto's moon, is almost as big as Pluto itself. Instead, scientists have created specific criteria to determine how an object is classed. However, as telescopes increase their range and computers process images with greater clarity, new information continually challenges the current understanding of the solar system.

As more distant bodies are discovered, better theories for their quantity and mass, their origins, and their relation to the rest of the solar system have been propounded. In 2005, a body bigger than Pluto was found and precipitated the argument: Was it the 10th planet or was, in fact, Pluto not even a planet itself? Because we have come to know that Pluto and its moon, Charon, orbit in a vast cloud of like objects, calling it a planet no longer made sense. And so, a new class of objects was born: the dwarf planets.

Every day, new data streams back to Earth from satellites and space missions. Early in 2004, scientists proved that standing liquid water once existed on Mars, just a month after a mission visited a comet and discovered that the material in its nucleus is as strong as some *rocks* and not the loose pile of ice and dust expected. The MESSENGER mission to Mercury, launched in 2004, has thus far completed three flybys and will enter Mercury orbit at 2011. The mission has already proven that Mercury's core is still molten, raising fundamental questions about processes of planetary evolution, and it has sent back to Earth intriguing information about the composition of Mercury's crust. Now the New Horizons mission is on its way to make the first visit to Pluto and the Kuiper belt. Information arrives from space observations and Earth-based experiments, and scientists attempt to explain what they see, producing a stream of new hypotheses about the formation and evolution of the solar system and all its parts.

The graph below shows the number of moons each planet has; large planets have more than small planets, and every year scientists discover new bodies orbiting the gas giant planets. Many bodies of substantial size orbit in the asteroid belt, or the Kuiper belt, and many sizable asteroids cross the orbits of planets as they make their way around the Sun. Some planets' moons are unstable and will in the near future (geologically speaking) make new ring systems as they crash into their hosts. Many moons, like Neptune's giant Triton, orbit their planets backward (clockwise when viewed from

The mass of the planet appears to control the number of moons it has; the large outer planets have more moons than the smaller inner planets.

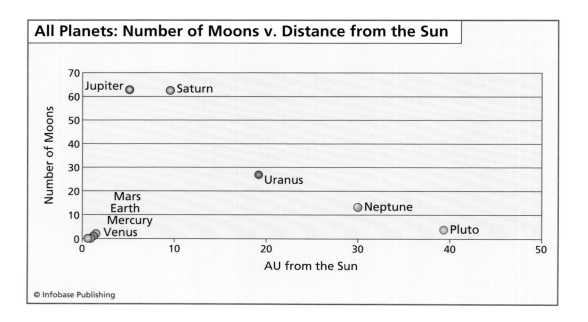

All Planets: Number of Moons v. Distance from the Sun

© Infobase Publishing

the North Pole, the opposite way that the planets orbit the Sun). Triton also has the coldest surface temperature of any moon or planet, including Pluto, which is much farther from the Sun. The solar system is made of bodies in a continuum of sizes and ages, and every rule of thumb has an exception.

Perhaps more important, the solar system is not a static place. It continues to evolve—note the drastic climate changes we are experiencing on Earth as just one example—and our ability to observe it continues to evolve, as well. Just five planets visible to the naked eye were known to ancient peoples: Mercury, Venus, Mars, Jupiter, and Saturn. The Romans gave these planets the names they are still known by today. Mercury was named after their god Mercury, the fleet-footed messenger of the gods, because the planet Mercury seems especially swift when viewed from Earth. Venus was named for the beautiful goddess Venus, brighter than anything in the sky except the Sun and Moon. The planet Mars appears red even from Earth and so was named after Mars, the god of war. Jupiter is named for the king of the gods, the biggest and most powerful of all, and Saturn was named for Jupiter's father. The ancient Chinese and the ancient Jews recognized the planets as well, and the Maya (250–900 C.E., Mexico and environs) and Aztec (~1100–1700 C.E., Mexico and environs) knew Venus by the name Quetzalcoatl, after their god of good and light, who eventually also became their god of war.

Science is often driven forward by the development of new technology, allowing researchers to make measurements that were previously impossible. The dawn of the new age in astronomy and study of the solar system occurred in 1608, when Hans Lippenshey, a Dutch eyeglass-maker, attached a lens to each end of a hollow tube and thus created the first telescope. Galileo Galilei, born in Pisa, Italy, in 1564, made his first telescope in 1609 from Lippenshey's model. Galileo soon discovered that Venus has phases like the Moon does and that Saturn appeared to have "handles." These were the edges of Saturn's rings, though the telescope was not strong enough to resolve the rings correctly. In 1610, Galileo discovered four of Jupiter's moons, which are still called the Galilean satellites. These four moons were the proof that not every heavenly body orbited the Earth as Ptolemy, a Greek philosopher, had

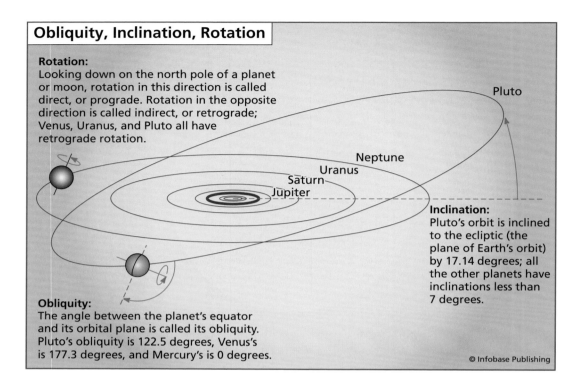

Obliquity, Inclination, Rotation

Rotation:
Looking down on the north pole of a planet or moon, rotation in this direction is called direct, or prograde. Rotation in the opposite direction is called indirect, or retrograde; Venus, Uranus, and Pluto all have retrograde rotation.

Pluto

Neptune
Uranus
Saturn
Jupiter

Inclination:
Pluto's orbit is inclined to the ecliptic (the plane of Earth's orbit) by 17.14 degrees; all the other planets have inclinations less than 7 degrees.

Obliquity:
The angle between the planet's equator and its orbital plane is called its obliquity. Pluto's obliquity is 122.5 degrees, Venus's is 177.3 degrees, and Mercury's is 0 degrees.

© Infobase Publishing

Obliquity, orbital inclination, and rotation direction are three physical measurements used to describe a rotating, orbiting body.

asserted around 140 C.E. Galileo's discovery was the beginning of the end of the strongly held belief that the Earth is the center of the solar system, as well as a beautiful example of a case where improved technology drove science forward.

The concept of the Earth-centered solar system is long gone, as is the notion that the heavenly spheres are unchanging and perfect. Looking down on the solar system from above the Sun's north pole, the planets orbiting the Sun can be seen to be orbiting counterclockwise, in the manner of the original *protoplanetary disk* of material from which they formed. (This is called *prograde* rotation.) This simple statement, though, is almost the end of generalities about the solar system. Some planets and dwarf planets spin backward compared to the Earth, other planets are tipped over, and others orbit outside the *ecliptic* plane by substantial angles, Pluto in particular (see the following figure on *obliquity* and orbital *inclination*). Some planets and moons are still hot enough to be volcanic, and some produce *silicate* lava (for example, the Earth and Jupiter's moon Io), while others have exotic lavas made of molten ices (for example, Neptune's moon Triton).

Today, we look outside our solar system and find planets orbiting other stars, more than 400 to date. Now our search for signs of life goes beyond Mars and Enceladus and Titan and reaches to other star systems. Most of the science presented in this set comes from the startlingly rapid developments of the last 100 years, brought about by technological development.

The rapid advances of planetary and heliospheric science and the astonishing plethora of images sent back by missions motivate the revised editions of the Solar System set. The multivolume set explores the vast and enigmatic Sun at the center of the solar system and moves out through the planets, dwarf planets, and minor bodies of the solar system, examining each and comparing them from the point of view of a planetary scientist. Space missions that produced critical data for the understanding of solar system bodies are introduced in each volume, and their data and images shown and discussed. The revised editions of *The Sun, Mercury, and Venus, The Earth and the Moon,* and *Mars* place emphasis on the areas of unknowns and the results of new space missions. The important fact that the solar system consists of a continuum of sizes and types of bodies is stressed in the revised edition of *Asteroids, Meteorites, and Comets.* This book discusses the roles of these small bodies as recorders of the formation of the solar system, as well as their threat as *impactors* of planets. In the revised edition of *Jupiter and Saturn,* the two largest planets are described and compared. In the revised edition of *Uranus, Neptune, Pluto, and the Outer Solar System,* Pluto is presented in its rightful, though complex, place as the second-largest known of a extensive population of icy bodies that reach far out toward the closest stars, in effect linking the solar system to the Galaxy itself.

This set hopes to change the familiar and archaic litany *Mercury, Venus, Earth, Mars, Jupiter, Saturn, Uranus, Neptune, Pluto* into a thorough understanding of the many sizes and types of bodies that orbit the Sun. Even a cursory study of each planet shows its uniqueness along with the great areas of knowledge that are unknown. These titles seek to make the familiar strange again.

Acknowledgments

Foremost, profound thanks to the following organizations for the great science and adventure they provide for humankind and, on a more prosaic note, for allowing the use of their images for these books: the National Aeronautics and Space Administration (NASA) and the National Oceanic and Atmospheric Administration (NOAA), in conjunction with the Jet Propulsion Laboratory (JPL) and Malin Space Science Systems (MSSS). A large number of missions and their teams have provided invaluable data and images, including the Solar and Heliospheric Observer (SOHO), Mars Global Surveyor (MGS), Mars Odyssey, the Mars Exploration Rovers (MERs), Galileo, Stardust, Near-Earth Asteroid Rendezvous (NEAR), and Cassini. Special thanks to Steele Hill, SOHO Media Specialist at NASA, who prepared a number of images from the SOHO mission, to the astronauts who took the photos found at Astronaut Photography of the Earth, and to the providers of the National Space Science Data Center, Great Images in NASA, and the NASA/JPL Planetary Photojournal, all available on the Web (addresses given in the reference section).

Many thanks also to Frank K. Darmstadt, executive editor; to Jodie Rhodes, literary agent; and to E. Marc Parmentier at Brown University for his generous support.

Introduction

*U*ranus, Neptune, Pluto, and the Outer Solar System, Revised Edition enters the farthest reaches of the solar system, including the distant gas planets Uranus and Neptune and the regions of asteroids and comets known as the Kuiper belt and the Oort cloud. These are the areas in the solar system that, in many ways, are the least known, and experiencing some of the fastest rates of new discovery. Unlike all the planets closer to the Sun, known since antiquity, the farthest reaches are the discoveries of the modern world: Uranus was discovered in 1781, Neptune in 1846, Pluto in 1930, the Kuiper belt group of objects in 1992, and though the Oort cloud has been theorized since 1950, its first member was just found in 2004. The discovery of the outer planets made such an impression on the minds of humankind that they were even immortalized in the names of newly discovered *elements:* uranium, neptunium, and plutonium, that astonishingly deadly constituent of atomic bombs.

Scientific theories rely on observations that produce data: temperatures, compositions, densities, sizes, times, patterns, or appearances. There is very little data on these outer solar system bodies, compared to what is known about Earth's neighbors the Moon and Mars, and even Jupiter and Saturn. In the cases of Neptune and Uranus, only the mission of *Voyager 2* during the mid-1980s attempted close observations. The extreme distance of these bodies from the Earth hinders Earth-based observations. Because there is so little data on Uranus, there are fewer scientists conducting research on the planet (they have little data for analysis and hypotheses testing). Much of the new science on this part of the solar system awaits new space missions to this region. The discrepancy in missions to the terrestrial planets compared to the outer planets is shown

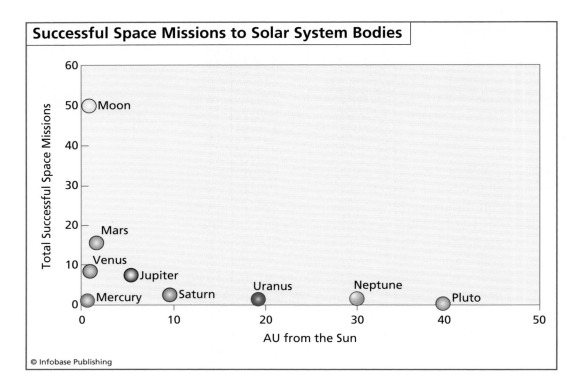

Successful Space Missions to Solar System Bodies

© Infobase Publishing

The approximate number of successful space missions from all nations to each of the planets and the Moon shows that the Moon is by far the most visited body, only Pluto has had no missions, and Mercury is as neglected as Uranus and Neptune. The definition of a successful mission is arguable, so totals for Mars and the Moon in particular may be disputed.

in the figure above; only Pluto has never had a space mission approach it.

Pluto and its neighbors will finally be visited by the *New Horizons* mission to the outer planets. The spacecraft, launched in 2006, will reach Pluto in 2015. This volume is updated with the state of knowledge of the outer reaches of the solar system, ready for the flood of new data to come from this mission.

Part One of the book discusses what data there is on the distant gas planets and investigates theories about their formation and evolution. All the gas giant planets, including Uranus and Neptune, are thought to have accumulated as masses of heterogeneous material. The small amount of very dense material available so far out in the nebular cloud of the early solar system fell through self-gravity into the center of each primordial planetary mass, forming whatever rocky or metallic *core* each planet might now have. The liquid and gaseous material that makes up the vast bulk of each planet forms layers according to its response to pressure and temperature. Though these planets have relatively low density, the heat of formation may still be influencing their interior circulation today.

Neptune and Uranus are twins in terms of size, internal structure, and color, but they differ in important ways. Uranus, for example, produces virtually no heat internally, while Neptune produces more heat relative to what it receives from the Sun than does any other planet. They both pose special challenges to theories of planet formation. These are huge planets, 15 and 17 times the mass of Earth, respectively, and yet they formed in the outer solar system where the density of material in the early solar system is thought to be very low and the orbits are huge, leading to less chance of collisions and collection of material. One theory is that both planets formed in the region of Jupiter and Saturn and were scattered outward in the solar system when Jupiter became huge and acquired its giant gas envelope.

Neptune and Uranus also have huge cores in relation to their overall planetary mass: probably 60 to 80 percent, as compared to Jupiter and Saturn's 3 to 15 percent. The question then is, how did huge cores form, and then fail to attract gravitationally the same fractions of gaseous atmospheres that Jupiter and Saturn did? One possibility, in contradiction to the idea that they formed near Jupiter and Saturn and then were thrown further out, is that Uranus and Neptune formed in the orbits they now inhabit. These planets' orbits are so huge and the protoplanetary disk so sparse at those distances from the Sun that the planetary cores would have taken a long time to form, much longer than Jupiter's and Saturn's.

Thus, very little is known about these remote gas giants, both because data is scarce and because scientists are still forming ideas about how they were created. Yet as remote as these two planets are, there is much remaining material farther out in the solar system still to be studied. Beyond the last two gas giant planets lie fascinating small bodies in closely spaced orbits, the home of Pluto and Charon, and the sources of long-period comets. Data on more distant and smaller objects is even more difficult to obtain. The fact that Pluto and its moon Charon are part of a much larger population of icy and rocky bodies now known as the Kuiper belt has been known for

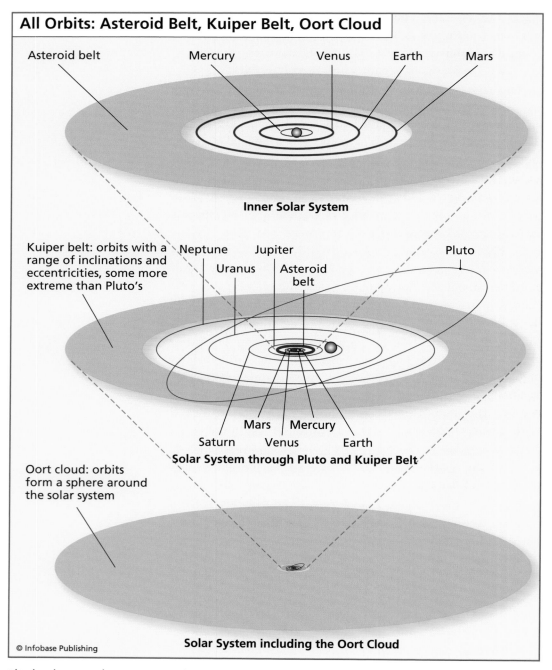

All Orbits: Asteroid Belt, Kuiper Belt, Oort Cloud

Asteroid belt   Mercury   Venus   Earth   Mars

Inner Solar System

Kuiper belt: orbits with a range of inclinations and eccentricities, some more extreme than Pluto's

Neptune   Jupiter   Pluto

Uranus   Asteroid belt

Mars   Mercury

Saturn   Venus   Earth

Solar System through Pluto and Kuiper Belt

Oort cloud: orbits form a sphere around the solar system

© Infobase Publishing

Solar System including the Oort Cloud

This book covers the outermost solar system, including Uranus, Neptune, Kuiper belt, and Oort cloud, whose orbits and ranges of orbits are highlighted here. All the orbits are far closer to circular than shown in this oblique view, which was chosen to show the inclination of Pluto's orbit to the ecliptic.

only about a decade. Pluto, Charon, and the other bodies that make up the Kuiper belt are the subject of Part Two of this book; their orbits in relation to those of the others covered in this book are shown in the figure on page xviii.

While awaiting the *New Horizons* data, the planetary science community has been working on theory. A critical step in understanding the early evolution of the solar system and the formation of the enigmatic Kuiper belt was begun in 2006 with the publication of a new model for solar system dynamics, the Nice Model, named after the town in France where the team met to develop the science. The details of this foundational work are described in a new chapter, "The Nice Model for Kuiper Belt Formation."

In 2004 the first body orbiting in the distant and until now theoretical Oort cloud was discovered. In Part Three, what is known about the Oort cloud is described, along with the reasoning that lead to its theorization and the technology that has allowed its actual discovery. The Kuiper belt and Oort cloud currently are under intense study. Although they are extremely distant and the bodies in them very small, new high-resolution imaging techniques are allowing scientists to discover many new objects in these regions, and they are beginning to better understand the populations at the edges of the solar system. This book describes what is now known about the bodies in these distant regions and how they interact with the inner solar system and also with stars entirely outside this solar system.

PART ONE

URANUS AND NEPTUNE

Uranus: Fast Facts about a Planet in Orbit

In the 1760s Johann Daniel Titius, a Prussian astronomer at the University of Wittenberg, began thinking about the distances of the planets from the Sun. Why are the inner planets closer together than are the outer planets? In 1766 he calculated the average distance that each planet lies from the Sun, and he noted that each planet is about 1.5 times farther from the Sun than the previous planet. This rule was published and made famous in 1772 (without attribution to Titius) by Johann Elert Bode, a German astronomer and director of the Berlin Observatory, in his popular book *Anleitung zur Kenntnis des gestirnten Himmels* (Instruction for the knowledge of the starry heavens). Bode did so much to publicize the law, in fact, that it is often known simply as Bode's Law. Both scientists noticed that the rule predicts that a planet should exist between Mars and Jupiter, though no planet was known to be there. There began a significant effort to find this missing planet, leading to the discovery of the main asteroid belt.

Johann Daniel Titius noticed in the mid-18th century that if the planets are numbered beginning with Mercury = 0, Venus = 3, and doubling thereafter, so Earth = 6, Mars = 12, and so on, and four is added to each of the planets' numbers, and then each is divided by 10, a series is created that very closely approximates the planet's distances from the Sun in *AU*. The final series he came up with is 0.4, 0.7, 1.0, 1.6, 2.8, 5.2, 10. In the table on the next page the planets are listed with their distances from the Sun and the Titius-Bode rule prediction. Remember that these calculations are in astronomical units: Mercury is on average 0.4 AU away from the Sun, and the Earth is 1 AU away.

The original formulation by Titius, published by Bode, stated mathematically, was

$$a = \frac{n + 4}{10},$$

where *a* is the average distance of the planet from the Sun in AU, and $n = 0, 3, 6, 12, 24, 48. \ldots$

Though now the rule is often written, with the same results, as

$$r = 0.4 + 0.3 \, (2^n),$$

where *r* is the orbital radius of the planet, and *n* is the number of the planet.

This series actually predicts where Uranus (pronounced "YOOR-un-us") was eventually found, though, sadly, it completely fails for Neptune and Pluto. There is no physical basis for the formation of this rule and there are still no good theories for why the rule works so well for the planets up through Uranus.

About 10 years later, on March 13, 1781, Friedrich Wilhelm Herschel, an amateur astronomer working in Bath, England, first recognized Uranus as a planet (born in Germany in 1738, he was known as Frederick William Herschel throughout his adult life in England, and later as Sir Frederick William Herschel). Galileo had seen Uranus in 1548, 233 years earlier, but did not recognize it as a new planet. Uranus is so dim and appears to move so slowly from the Earth's vantage point that it is easily mistaken for a star. In 1690 John Flamsteed, the British Royal Astronomer, also saw Uranus, and also failed to recognize it as a planet. In fact, from the notes of these and other astronomers, it is known that Uranus was seen and

TITIUS-BODE RULE PREDICTIONS

Planet/belt	Average AU from the Sun	Titius-Bode rule prediction
Mercury	0.387	0.4
Venus	0.723	0.7
Earth	1.0	1.0
Mars	1.524	1.6
Asteroid Belt	2.77	2.8
Jupiter	5.203	5.2
Saturn	9.539	10.0
Uranus	19.18	19.6
Neptune	30.06	38.8
Pluto	39.44	77.2

described at least 21 times before 1781, but not once recognized as a planet.

Herschel was primarily a musician and composer, though music led him to astronomy (there are intriguing parallels between mathematics, astronomy, and music). In 1773 he bought his first telescope. He and his sister Caroline shared a house in England where they taught music lessons, and soon they were manufacturing telescopes in every moment of their spare time. Around 1770 Herschel decided to "review" the whole sky, that is, to sweep large areas and examine small areas in great detail, to gain "a knowledge of the construction of the heavens." He worked on this immense task for decades. On Tuesday, March 13, 1781, he wrote, "Between ten and eleven in the evening, while I was examining the small stars in the neighborhood of H Geminorum, I perceived one that appeared visibly larger than the rest." This was Uranus. Even Herschel first thought Uranus was a comet, though he wrote that "it appears as a ball instead of a point when magnified by a telescope, and shines steadily instead of twinkling." He wrote a short text called "Account of a Comet," which he submitted to the Royal Society of London. When it was finally determined through continued study of the body by Herschel and others that it was in fact a planet, Herschel's text proved he was its first discoverer.

The recognition of Uranus as a planet doubled the known size of the solar system, and meant that Herschel was the first discoverer of a planet since prehistoric times. Herschel named the planet Georgian Sidus, meaning George's star, after George III, king of England (who had recently had the very great disappointment of losing the American colonies). The planet was called Georgian Sidus in England for almost 70 years despite King George's relative unpopularity. Joseph-Jérôme Lefrançais de Lalande, a French astronomer and mathematician, named the planet Herschel, and in France they called it so for many decades. Other names were suggested, including Cybele, after the mythical wife of the god Saturn. Finally, Johann Elert Bode named the planet Uranus for the Greek and Roman god of the sky who was the father of Saturn, and that is the name that stuck (it also makes a nice symmetry: In mythology, Uranus is Saturn's father, Saturn is Jupiter's father, and Jupiter is the father of Mercury, Venus, and Mars).

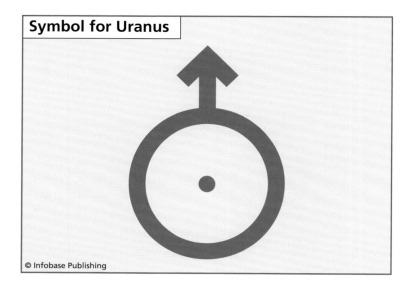

Symbol for Uranus

© Infobase Publishing

Many solar system objects have simple symbols; this is the symbol for Uranus.

Herschel went on to discover two moons of Uranus and two of Saturn. He died at the age of 84, the exact number of years in Uranus's orbit around the Sun, and so when he died Uranus was in the same place in the sky that it had been when he was born. His sister Caroline, his partner in all his astronomical works, may have been the first female professional astronomer. She made comet-hunting her specialty, and by the time of her death at 98, she had discovered eight new comets.

Each planet and some other bodies in the solar system (the Sun and certain asteroids) have been given its own symbol as a shorthand in scientific writing. The symbol for Uranus is shown in the figure on page 6.

A planet's rotation prevents it from being a perfect sphere. Spinning around an axis creates forces that cause the planet to swell at the equator and flatten slightly at the poles. Planets are thus shapes called oblate spheroids, meaning that they have different equatorial radii and polar radii. If the planet's equatorial radius is called r_e, and its polar radius is called r_p, then its flattening (more commonly called ellipticity, e, shown in the figures on pages 9 and 10) is defined as

$$e = \frac{r_e - r_p}{r_e}.$$

The larger radius, the equatorial, is also called the *semimajor axis*, and the polar radius is called the *semiminor axis*. The Earth's semi-major axis is 3,960.8 miles (6,378.14 km), and its semiminor axis is 3,947.5 miles (6,356.75 km), so its ellipticity is

$$e = \frac{3960.8 - 3947.5}{3960.8} = 0.00335.$$

Because every planet's equatorial radius is longer than its polar radius, the surface of the planet at its equator is farther from the planet's center than the surface of the planet at the poles. What effect does the mass have? Mass pulls with gravity (for more information on gravity, see the sidebar "What Makes Gravity?" on page 11). At the equator, where the radius of the planet is larger and the amount of mass beneath them is

FUNDAMENTAL INFORMATION ABOUT URANUS

Uranus is the third of the gas giant planets and shares many characteristics with the other gas planets, most markedly with Neptune. In 1664 Giovanni Cassini, an Italian astronomer, made an important observation of Jupiter: It is flattened at its poles and bulges at its equator. Cassini was exactly right in this observation, and though the effect is extreme on Jupiter, in fact all planets are slightly flattened. Gas planets are particularly susceptible to flattening, and Uranus is no exception, having a radius at its equator about 2 percent longer than its radius at its poles. Its mass is only about 5 percent of Jupiter's, but Uranus is less dense. Uranus thus becomes the third-largest planet, but because of its low density, it is the fourth most massive. These and other physical parameters for Uranus are given in this table.

FUNDAMENTAL FACTS ABOUT URANUS	
equatorial radius at the height where atmospheric pressure is one bar	15,882 miles (25,559 km), or four times Earth's radius
polar radius	15,518 miles (24,973 km)
ellipticity	0.0229, meaning the planet's equator is about 2 percent longer than its polar radius
volume	1.42×10^{13} cubic miles (5.914×10^{13} km^3), or 52 times Earth's volume
mass	1.91×10^{26} pounds (8.68×10^{25} kg), or 14.5 times Earth's mass
average density	79.4 pounds per cubic foot (1,270 kg/m^3), or 0.24 times Earth's density
acceleration of gravity on the surface at the equator	28.5 feet per squared seconds (8.69 m/sec^2), or 0.89 times Earth's gravity
magnetic field strength at the surface	2×10^{-5} tesla, similar to Earth's magnetic field
rings	11
moons	27 presently known

Ellipticity

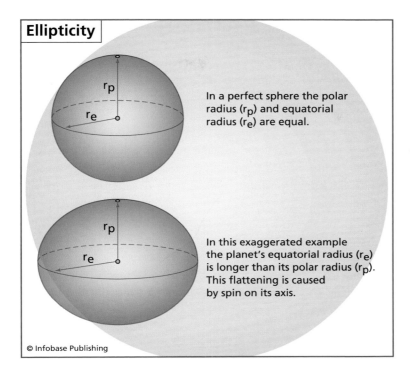

In a perfect sphere the polar radius (r_p) and equatorial radius (r_e) are equal.

In this exaggerated example the planet's equatorial radius (r_e) is longer than its polar radius (r_p). This flattening is caused by spin on its axis.

© Infobase Publishing

Ellipticity is the measure of by how much a planet's shape deviates from a sphere.

relatively larger, the pull of gravity is actually stronger than it is at the poles. Gravity is actually not a perfect constant on any planet: Variations in radius, topography, and the density of the material underneath make the gravity vary slightly over the surface. This is why planetary gravitational accelerations are generally given as an average value on the planet's equator.

Most of the planets orbit in almost exactly the same plane, and most of the planets rotate around axes that lie close to perpendicular to that plane. Though Uranus's orbit lies very close to the solar system's ecliptic plane, its equator lies at almost 98 degrees from its orbital plane. The angle between a planet's equatorial plane and the plane of its orbit is known as its obliquity (obliquity is shown for the Earth on page 12 and given in the table on page 13). The Earth's obliquity, 23.45 degrees, is intermediate in the range of solar system values. The planet with the most extreme obliquity is Venus, with an obliquity of 177.3 degrees, followed by Pluto, with an obliquity of 119.6 degrees. Obliquities above 90 degrees mean that the planet's north pole has passed through its orbital plane and now points south. This is similar to Uranus's state, with a rotational axis tipped until it almost lies flat in its orbital plane.

The only theory that seems to account for Uranus's great obliquity is that early in its accretionary history it was struck by another *planetesimal,* one almost the same size as the proto-Uranus, and the energy of this collision tipped over the rotating planet. Uranus's extreme obliquity, combined with its 84-year *orbital period,* means that its south and north poles are each in darkness for about 42 years at a time. Uranus's orbit is also anomalous: It is not exactly as it should be as predicted by theory. The differences in its orbit are caused by the gravity of Neptune, and these discrepancies predicted the existence of Neptune and led to its discovery. More measurements of Uranus's orbit are given in the table on page 13. For a complete description of the orbital elements, see chapter 5, "Neptune: Fast Facts about a Planet in Orbit."

The ellipticities of the planets differ largely as a function of their composition's ability to flow in response to rotational forces.

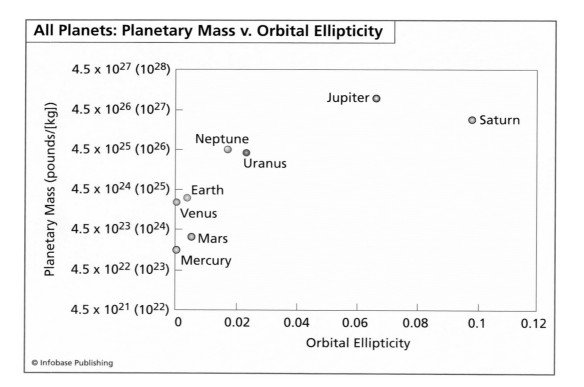

All Planets: Planetary Mass v. Orbital Ellipticity

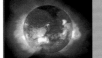

WHAT MAKES GRAVITY?

Gravity is among the least understood forces in nature. It is a fundamental attraction between all matter, but it is also a very weak force: The gravitational attraction of objects smaller than planets and moons is so weak that electrical or magnetic forces can easily oppose it. At the moment about the best that can be done with gravity is to describe its action: How much mass creates how much gravity? The question of what makes gravity itself is unanswered. This is part of the aim of a branch of mathematics and physics called string theory: to explain the relationships among the natural forces and to explain what they are in a fundamental way.

Sir Isaac Newton, the English physicist and mathematician who founded many of today's theories back in the mid-17th century, was the first to develop and record universal rules of gravitation. There is a legend that he was hit on the head by a falling apple while sitting under a tree thinking, and the fall of the apple under the force of Earth's gravity inspired him to think of matter attracting matter.

The most fundamental description of gravity is written in this way:

$$F = \frac{Gm_1 m_2}{r^2},$$

where F is the force of gravity, G is the universal gravitational constant (equal to 6.67×10^{-11} Nm^2/kg^2), m_1 and m_2 are the masses of the two objects that are attracting each other with gravity, and r is the distance between the two objects. (N is the abbreviation for newtons, a metric unit of force.)

Immediately, it is apparent that the larger the masses, the larger the force of gravity. In addition, the closer together they are, the stronger the force of gravity, and because r is squared in the denominator, gravity diminishes very quickly as the distance between the objects increases. By substituting numbers for the mass of the Earth (5.9742×10^{24} kg), the mass of the Sun (1.989×10^{30} kg), and the distance between them, the force of gravity between the Earth and Sun is shown to be 8×10^{21} pounds per foot (3.56×10^{22} N). This is the force that keeps the Earth in orbit around the Sun. By comparison, the force of gravity between a piano player and her piano when she sits playing is about 6×10^{-7} pounds per foot (2.67×10^{-6} N). The force of a pencil pressing down in the palm of a hand under the influence of Earth's gravity is about 20,000 times stronger than the gravitational attraction between the player and the piano! So, although the player and the piano are attracted to each other by gravity, their masses are so small that the force is completely unimportant.

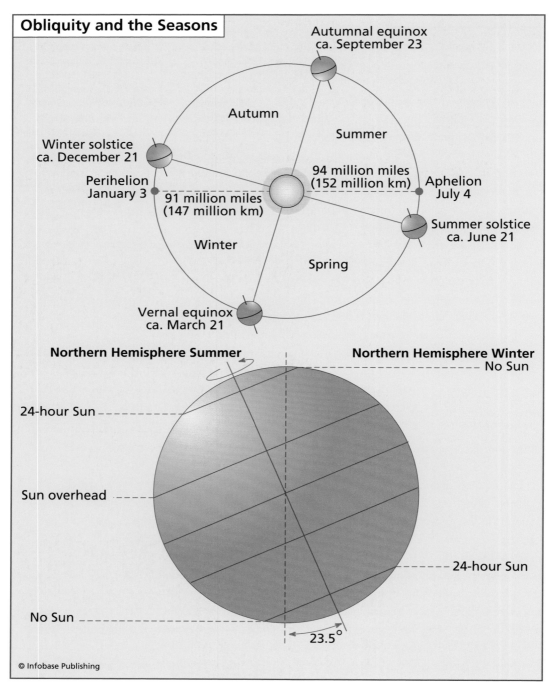

Obliquity and the Seasons

Autumnal equinox
ca. September 23

Autumn

Summer

Winter solstice
ca. December 21

94 million miles
(152 million km)

Aphelion
July 4

Perihelion
January 3

91 million miles
(147 million km)

Summer solstice
ca. June 21

Winter

Spring

Vernal equinox
ca. March 21

Northern Hemisphere Summer

Northern Hemisphere Winter
No Sun

24-hour Sun

Sun overhead

24-hour Sun

No Sun

23.5°

© Infobase Publishing

A planet's obliquity (the inclination of its equator to its orbital plane) is the primary cause of seasons. This figure describes the obliquity of the Earth.

URANUS'S ORBIT	
rotation on its axis ("day")	17 Earth hours, 42 minutes
rotation speed at equator	5,788 miles per hour (9,315 km/hour)
rotation direction	*retrograde* (clockwise when viewed from above its North Pole); this means it spins backward compared to the Earth
sidereal period ("year")	83.74 Earth years
orbital velocity (average)	4.24 miles per second (6.83 km/sec)
sunlight travel time (average)	two hours, 39 minutes, and 31 seconds to reach Uranus
average distance from the Sun	1,783,939,400 miles (2,870,972,200 km), or 19.191 AU
perihelion	1,699,800,000 miles (2,735,560,000 km), or 18.286 AU from the Sun
aphelion	1,868,080,000 miles (3,006,390,000 km), or 20.096 AU from the Sun
orbital eccentricity	0.04717
orbital inclination to the ecliptic	0.77 degrees
obliquity (inclination of equator to orbit)	97.86 degrees

There is still an argument, however, over which of Uranus's poles is its north pole. Normally the north pole is easily designated as the one that is above the ecliptic, but Uranus's poles lie so close to the ecliptic that it is not clear which should be called the north pole. The magnetic field does not help either, because the orientation of a planet's magnetic field is believed to change over time, as has the Earth's. The direction of the magnetic field is especially unhelpful in the case of Uranus, which has a magnetic field that is not remotely aligned with its poles. If in fact the planet has tipped over slightly less than 90 degrees, rather than slightly more, as described here, then its rotation is *direct* (in the same sense as the Earth's), rather than retrograde, as listed here.

The Interior of Uranus

Saturn and Jupiter consist mainly of hydrogen and helium, the two most common elements in the solar system. Since material in the solar nebula is thought to become less dense with distance from the Sun, planets exterior to Saturn and Jupiter should have more hydrogen and helium. Uranus, however, has far more heavy elements than do Jupiter and Saturn. This unexpected result is simply the first of the paradoxes of this planet. Uranus's orbital axis lies almost flat in its orbital plane, for reasons that are not fully understood. Uranus also produces no detectable heat from its interior. Without heat release, there is little information to be had about the internal temperature profile of the planet. Uranus and its neighbor Neptune remain largely unknown.

COMPOSITION

The outer portions of Uranus consist of gaseous helium and hydrogen, like its other neighbors in the outer solar system, but Uranus also has methane (CH_4), ammonia (NH_3), and water (H_2O) in its atmosphere (for more, see the sidebar "Elements and Isotopes" on page 20). The methane, though it makes up less than 3 percent of the atmosphere, absorbs enough red light to give the planet its distinctive blue-green appearance: By absorbing most of the red wavelengths, mainly blue and green wavelengths remain to be reflected. The bulk of the planet is made up of a range of ices, and the gaseous helium and hydrogen together make up only 15 to 20 percent of the planet, far less than Jupiter and Saturn are thought to have. If Uranus has a rocky or icy core, it must be on the order of a few Earth masses or less.

Elements heavier than helium and hydrogen make up 80 to 85 percent of Uranus and Neptune by mass. In fact, both Uranus and Neptune have far more of these heavy elements (oxygen, carbon, and heavier elements) than either Jupiter or Saturn do. Because Jupiter and Saturn are closer to the Sun, they would normally be expected to have heavier constituents than planets farther out. This paradox is not well explained. One theory is that both Uranus and Neptune originally formed closer to the Sun, but were thrown further out by the gravitational perturbations of the giant, growing Jupiter, but this is a tenuous theory and the heavy elements in Uranus and Neptune remain unexplained.

The carbon in these planets combines with hydrogen to form methane, and oxygen combines with hydrogen to form water. Both these molecules are heavier than the gases around them in the planets' outer atmospheres, and both these molecules will condense in the cold temperatures present there, and sink into the planet's interior. As pressure increases with depth in the planet, the gases become liquid (for more on pressure, see the sidebar "What Is Pressure?" on page 17). This liquid may be largely water. Below this is hypothesized to be a layer of partly solid ammonia and methane, followed by the planet's rocky or icy core, which itself is probably about the mass of Earth. So little about Uranus's core is known that its density cannot be constrained to a smaller range than 283 to 566 lb/ft^3 (4,500 to 9,000 kg/m^3), depending on whether there is a molten rocky core in addition to the solid rocky core, and depending on what the rest of the density distributions are within Uranus.

There are theories of formation for the gas giant planets but many unanswered questions as well. Scientists assume that the solar nebula was cooler farther from the Sun, and therefore more *volatile* compounds could condense into gases and liquids and form planets. Because each of these outer planets has a huge orbit from which to scavenge nebular material, the planets could grow very large. Uranus and Neptune both seem to have larger cores in relation to their outer diameters than Jupiter or Saturn have.

Very early in its formation, possibly before nuclear fusion began, the Sun went through a special stage in its evolution called the T-Tauri stage. T-Tauri refers to a class of protostars, generally less than about 2.5 times the mass of the Sun, before fusion begins; the stars shine only from heat of gravitational collapse. During this stage the protostar again emits giant jets of charged particles from its north and south poles. These jets are large and hot, as long as several hundred AU.

When the T-Tauri stage of stellar evolution was first discovered, scientists thought that the protostar emitted radiation and material jets in the plane of its protoplanetary disk, that is, straight toward and past its young accreting planets. The T-Tauri winds were called upon again and again by planetary scientists to strip the early atmospheres and surfaces off the young planets; that was the explanation for the chemically evolved atmosphere of the Earth, as well as Mercury's lack of an atmosphere: It has been removed by the T-Tauri stage winds. This was also a suggested explanation for the loss of gases from Neptune and Uranus.

Soon after, astronomers discovered that the T-Tauri winds extend from the poles of the protostar and not in the plane of the planets. Planetary geology text, however, lagged behind this knowledge. To this day some texts call on T-Tauri winds to strip the early atmospheres from young planets and to remove the last dust and gas from the disk. Now it is known, however, that T-Tauri winds do not flow past the growing planets, but also that planets are unlikely to have accreted by the time the T-Tauri stage occurs. Planet building is just beginning, if at all, when the protostar undergoes this stage. Other hypotheses must be sought to explain the large rocky cores of Neptune and Uranus.

WHAT IS PRESSURE?

The simple definition of pressure *(p)* is that it is force *(F)* per area *(a)*:

$$p = \frac{F}{a}.$$

Atmospheric pressure is the most familiar kind of pressure and will be discussed below. Pressure, though, is something felt and witnessed all the time, whenever there is a force being exerted on something. For example, the pressure that a woman's high heel exerts on the foot of a person she stands on is a force (her body being pulled down by Earth's gravity) over an area (the area of the bottom of her heel). The pressure exerted by her heel can be estimated by calculating the force she is exerting with her body in Earth's gravity (which is her weight, here guessed at 130 pounds, or 59 kg, times Earth's gravitational acceleration, 32 ft/sec², or 9.8 m/sec²) and dividing by the area of the bottom of the high heel (here estimated as one square centimeter):

$$p = \frac{\left(59 kg\right)\left(9.8 \text{ m/sec}^2\right)}{\left(0.01^2 m^2\right)} = 5,782,000 \; kg \, / \, m\sec^2$$

The resulting unit, kg/ms², is the same as N/m and is also known as the pascal (Pa), the standard unit of pressure (see appendix 1, "Units and Measurements," to understand more). Although here pressure is calculated in terms of pascals, many scientists refer to pressure in terms of a unit called the atmosphere. This is a sensible unit because one atmosphere is approximately the pressure felt from Earth's atmosphere at sea level, though of course weather patterns cause continuous fluctuation. (This fluctuation is why weather forecasters say "the barometer is falling" or "the barometer is rising": The measurement of air pressure in that particular place is changing in response to moving masses of air, and these changes help indicate the weather that is to come.) There are about 100,000 pascals in an atmosphere, so the pressure of the woman's high heel is about the same as 57.8 times atmospheric pressure.

What is atmospheric pressure, and what causes it? Atmospheric pressure is the force the atmosphere exerts by being pulled down toward the planet by the planet's

(continues)

(continued)

gravity, per unit area. As creatures of the Earth's surface, human beings do not notice the pressure of the atmosphere until it changes; for example, when a person's ears pop during a plane ride because the atmospheric pressure lessens with altitude. The atmosphere is thickest (densest) at the planet's surface and gradually becomes thinner (less and less dense) with height above the planet's surface. There is no clear break between the atmosphere and space: the atmosphere just gets thinner and thinner and less and less detectable. Therefore, atmospheric pressure is greatest at the planet's surface and becomes less and less as the height above the planet increases. When the decreasing density of the atmosphere and gravity are taken into consideration, it turns out that atmospheric pressure decreases exponentially with altitude according to the following equation:

$$p(z) = p_o e^{-\alpha z},$$

where $p(z)$ is the atmospheric pressure at some height above the surface z, p_o is the pressure at the surface of the planet, and α is a number that is constant for each planet and is calculated as follows:

$$\alpha = \frac{g\rho_0}{p_0},$$

where g is the gravitational acceleration of that planet, and ρ_o is the density of the atmosphere at the planet's surface.

   Just as pressure diminishes in the atmosphere from the surface of a planet up into space, pressure inside the planet increases with depth. Pressure inside a planet can be approximated simply as the product of the weight of the column of solid material above the point in question and the gravitational acceleration of the planet. In other words, the pressure P an observer would feel if he or she were inside the planet is caused by the weight of the material over the observer's head (approximated as ρh, with h the depth you are beneath the surface and ρ the density of the material between the observer and the surface) being pulled toward the center of the planet by its gravity g:

$$P = \rho g h.$$

The deeper into the planet, the higher the pressure.

INTERNAL TEMPERATURES

Uranus's rotation axis (which lies almost in its orbital plane) remains in a fixed direction as the planet orbits around the Sun. As a result, for one-quarter of its orbit the Sun shines more or less directly on its north pole (and the south pole remains in complete darkness), and then for the next quarter, the Sun shines on the planet's rotating equator (which then experiences day and night), and for the next quarter, on its south pole, and then back on the equator. Each of these quarter orbits lasts about 41 years.

When *Voyager 2* visited Uranus in 1986, the planet's south pole was pointing almost directly at the Sun, and therefore the planet was experiencing one of its most extreme temperature gradients from pole to pole. In spite of these long, focused periods of heating, which are much more extreme than those of the seasons on Earth, temperature fluctuations are small. Uranus seems to be able to transport heat very efficiently through its atmosphere, equalizing the effects of very different heat inputs on different sides of the planet. The measurements of *Voyager 2* indicated that, if anything, Uranus is slightly hotter at its equator than at either of its poles.

Unlike Jupiter and Saturn, Uranus does not give off more heat than it receives from the Sun. Its large solid core does not produce the energy that the larger, more gaseous planets do. Uranus's heat flux through its surface is another anomaly that sets it apart from other planets: The other planets leak heat through their surfaces, releasing energy from internal *radioactive* decay or other processes (for more, see the sidebar "Elements and Isotopes" on page 20). Heat loss through Uranus's surface, however, is undetectable. The planet seems to be losing no heat at all. Measuring heat loss from the interior of a planet is the best, and often the only, way to calculate its internal temperatures. Because Uranus is not losing any heat, its internal temperatures are largely unknown. If Uranus's interior temperatures are at all similar to the other gas giant planets, however, its interior may consist largely of a water ocean. Uranus and Neptune may then be other places in the solar system to look for life.

ELEMENTS AND ISOTOPES

All the materials in the solar system are made of *atoms* or of parts of atoms. A family of atoms that all have the same number of positively charged particles in their nuclei (the center of the atom) is called an *element:* Oxygen and iron are elements, as are aluminum, helium, carbon, silicon, platinum, gold, hydrogen, and well over 200 others. Every single atom of oxygen has eight positively charged particles, called protons, in its nucleus. The number of protons in an atom's nucleus is called its *atomic number:* All oxygen atoms have an atomic number of 8, and that is what makes them all oxygen atoms.

Naturally occurring nonradioactive oxygen, however, can have either eight, nine, or 10 uncharged particles, called neutrons, in its nucleus, as well. Different weights of the same element caused by addition of neutrons are called *isotopes.* The sum of the protons and neutrons in an atom's nucleus is called its *mass number.* Oxygen can have mass numbers of 16 (eight positively charged particles and eight uncharged particles), 17 (eight protons and nine neutrons), or 18 (eight protons and 10 neutrons). These isotopes are written as ^{16}O, ^{17}O, and ^{18}O. The first, ^{16}O, is by far the most common of the three isotopes of oxygen.

Atoms, regardless of their isotope, combine together to make molecules and compounds. For example, carbon (C) and hydrogen (H) molecules combine to make methane, a common gas constituent of the outer planets. Methane consists of one carbon atom and four hydrogen atoms and is shown symbolically as CH_4. Whenever a subscript is placed by the symbol of an element, it indicates how many of those atoms go into the makeup of that molecule or compound.

Quantities of elements in the various planets and moons, and ratios of isotopes, are important ways to determine whether the planets and moons formed from the same material or different materials. Oxygen again is a good example. If quantities of each of the oxygen isotopes are measured in every rock on Earth and a graph is made of the ratios of $^{17}O/^{16}O$ versus $^{18}O/^{16}O$, the points on the graph will form a line with a certain slope (the slope is 1/2, in fact). The fact that the data forms a line means that the material that formed the Earth was homogeneous; beyond rocks, the oxygen isotopes in every living thing and in the atmosphere also lie on this slope. The materials on the Moon also show this same slope. By measuring oxygen isotopes in many different kinds of solar system materials, it has now been shown that the slope of the plot $^{17}O/^{16}O$ versus $^{18}O/^{16}O$ is one-half for every object, but each object's line is offset from the others by some amount. Each solar system object lies along a different parallel line.

At first it was thought that the distribution of oxygen isotopes in the solar system was determined by their mass: The more massive isotopes stayed closer to the huge gravitational force of the Sun, and the lighter isotopes strayed farther out into the solar system. Studies of very primitive meteorites called chondrites, thought to be the most primitive, early material in the solar system, showed to the contrary that they have heterogeneous oxygen isotope ratios, and therefore oxygen isotopes were not evenly spread in the early solar system. Scientists then recognized that temperature also affects oxygen isotopic ratios: At different temperatures, different ratios of oxygen isotopes condense. As material in the early solar system cooled, it is thought that first aluminum oxide condensed, at a temperature of about 2,440°F (1,340°C), and then calcium-titanium oxide $(CaTiO_3)$, at a temperature of about 2,300°F (1,260°C), and then a calcium-aluminum-silicon-oxide $(Ca_2Al_2SiO_7)$, at a temperature of about 2,200°F (1,210°C), and so on through other compounds down to iron-nickel alloy at 1,800°F (990°C) and water, at −165°F (−110°C) (this low temperature for the condensation of water is caused by the very low pressure of space). Since oxygen isotopic ratios vary with temperature, each of these oxides would have a slightly different isotopic ratio, even if they came from the same place in the solar system.

The key process that determines the oxygen isotopes available at different points in the early solar system nebula seems to be that simple compounds created with ^{18}O are relatively stable at high temperatures, while those made with the other two isotopes break down more easily and at lower temperatures. Some scientists therefore think that ^{17}O and ^{18}O were concentrated in the middle of the nebular cloud, and ^{16}O was more common at the edge. Despite these details, though, the basic fact remains true: Each solar system body has its own slope on the graph of oxygen isotope ratios.

Most atoms are stable. A carbon-12 atom, for example, remains a carbon-12 atom forever, and an oxygen-16 atom remains an oxygen-16 atom forever, but certain atoms eventually disintegrate into a totally new atom. These atoms are said to be "unstable"

(continues)

(continued)

or "radioactive." An unstable atom has excess internal energy, with the result that the nucleus can undergo a spontaneous change toward a more stable form. This is called "radioactive decay." Unstable isotopes (radioactive isotopes) are called "radioisotopes." Some elements, such as uranium, have no stable isotopes. The rate at which unstable elements decay is measured as a "half-life," the time it takes for half of the unstable atoms to have decayed. After one half-life, half the unstable atoms remain; after two half-lives, one-quarter remain, and so forth. Half-lives vary from parts of a second to millions of years, depending on the atom being considered. Whenever an isotope decays, it gives off energy, which can heat and also damage the material around it. Decay of radioisotopes is a major source of the internal heat of the Earth today: The heat generated by accreting the Earth out of smaller bodies and the heat generated by the giant impactor that formed the Moon have long since conducted away into space.

MAGNETIC FIELD

Uranus's magnetic field has a north pole and a south pole, as does the Earth's magnetic field. This basic formation with two poles is known as a dipolar field. Uranus's dipolar magnetic pole is oriented 60 degrees from the planet's axis, and the center of its *dipole* is offset from the center of the planet toward the north pole by more than a third of the planet's radius! These anomalies indicate that the magnetic field may be created by electrical currents in shallow liquids and ices, rather than by a core dynamo. These currents create a magnetic dynamo, much as currents in the outer, liquid iron core on Earth produce its magnetic dynamo: The moving fluid iron in Earth's outer core acts as electrical currents, and every electrical current has an associated, enveloping magnetic field. The combination of convective currents and a spinning inner and outer core is called the dynamo effect. If the fluid motion is fast, large, and conductive enough, then a magnetic field can be not only created but also carried and deformed by the moving fluid. On Uranus, apparently, the flowing currents of ices act as an electrical current. At the surface, its magnetic field is about the same strength as the Earth's.

Uranus emits radio waves with frequencies in the 100 to 1,000 kHz range, emitted both in bursts and continuously. Uranus's radio wave emissions are more similar to Neptune's than to other planets'. These emissions are also oriented and rotate with the planet; at present, the strongest emissions are aimed away from the Earth. They seem to originate in the upper atmosphere along magnetic and auroral field lines and are thought to originate from electrons spiraling along field lines.

Research by Sabine Stanley and Jeremy Bloxham at Harvard University is beginning to shed light on the strange patterns of Uranus's magnetic field. Their modeling efforts, conducted on exceptionally fast and powerful computers, indicate that Uranus's magnetic field can be produced in thin layers of electrically conducting, convecting fluid in Uranus's interior. These sophisticated computer programs can reproduce both the pattern and strength of Uranus's magnetic field, as well as its angle and offset from the planet's rotational axis. These encouraging results show that theory can advance understanding of the planet even in the absence of new data. There is hope, then, that some of the other mysteries of Uranus's interior may begin to yield without new space missions to the planet, since none are planned at the current time.

3

Surface Appearance and Conditions on Uranus

Uranus can no longer be thought of as a featureless blue-green sphere. A number of problems have inhibited scientists' understanding of the surface conditions of the planet: First, instruments have not been sensitive enough to detect subtle clouds. Second, Uranus's lack of surface heat flow has made scientists believe that violent weather is unlikely. Finally, Uranus's seasons are so long that humankind has only observed the planet closely during one season. As observational techniques have improved, Uranus's weather is revealing its complexity. Uranus's atmosphere is about 83 percent hydrogen, 15 percent helium, and 2 percent methane (though Uranus has relatively little hydrogen in bulk; it is simply concentrated in the atmosphere). Methane molecules absorb red light and reflect or emit only blue and green light, so the methane in the outer layers of the planet gives the planet its bright blue-green color.

The temperature at Uranus's *tropopause* (the point where temperature stops falling with increasing altitude and actually begins to rise) is about −364°F (−220°C) and is near the pressure level of 0.1 bar (see figure on page 25). The clouds are found about 30 miles (50 km) below the tropopause, at a pressure of about one bar and a temperature of −328°F (−200°C). Deeper in Uranus, where the pressure is about three bars, the temperature rises to −243°F (−153°C). Above the Uranian tropopause the concentrations of methane and hydrogen decrease dramatically. The region above the tropopause is called the *stratosphere,* a stable layer of atmosphere on all planets, because temperature increases with altitude. Increasing temperature creates stability because warmer material is less dense than colder material, so naturally it stays buoyantly above the colder and therefore denser material. In Uranus's stratosphere the temperature rises from the low of about −364°F (−220°C) at the tropopause to a high of perhaps 890°F (480°C) in the *exosphere,* the uppermost layer of its atmosphere. The extreme warmth of the Uranian upper atmosphere is not well understood; there is no explanation from the weak solar heating at the distance of Uranus or from the immeasurably small energy output from the planet's interior. The other gas giant planets have similarly hot exospheres, and on Jupiter the heat is thought to come from ionizing reactions tied to Jupiter's magnetic field and its auroras. Perhaps a similar process is at work on Uranus and Neptune.

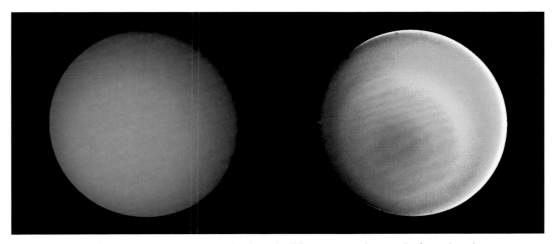

The picture at left shows Uranus in natural color. The blue-green color results from the absorption of red light by methane gas in Uranus's deep, cold, and remarkably clear atmosphere. The picture at right uses false color and extreme contrast enhancement to bring out subtle details in the polar region of Uranus, revealing a dark polar hood surrounded by a series of progressively lighter concentric bands. The bright orange and yellow strip at the lower edge of the planet's limb is an artifact of the image enhancement. (NASA/JPL)

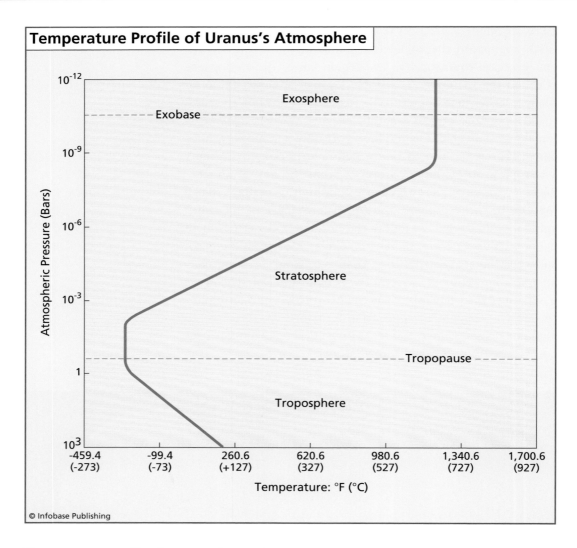

Temperature Profile of Uranus's Atmosphere

Exosphere

Exobase

Stratosphere

Tropopause

Troposphere

Atmospheric Pressure (Bars)

Temperature: °F (°C)

© Infobase Publishing

The temperature profile of the Uranian atmosphere is similar to those of the other gas giant planets.

What methane there is in the outer atmosphere freezes and settles back into the lower atmosphere with relatively little remixing. There are therefore few hydrocarbons for solar radiation to turn into smog, and the Uranian outer atmosphere remains clear. Because the outer atmosphere is both clear and warm, it has expanded much farther than the outer atmospheres of other gas giants. Even at the altitudes at which Uranus's rings orbit (1.64 to two times Uranus's radius) there is enough atmospheric gas to create drag on the rings. The smallest particles have the greatest surface-drag-to-mass ratio and so are most affected by the high atmosphere. The smallest particles are preferentially slowed by atmospheric drag, and they are the first to fall back into the tropopause, having been slowed to the point that Uranus's gravity pulls them in. This process tends to make Uranus's rings consist of larger particles.

From the images of *Voyager 2* and later observations it has long been thought that Uranus was almost featureless and completely lacked weather systems. It also stood to reason that if there was no measurable heat flux out of the planet and surface temperatures remained constant, then there was no driving force for winds. Recently, though, researchers have begun using large radar arrays on Earth to make high-resolution radar maps of Uranus (radar stands for radio detection and ranging, a technique that bounces radiation at radio wavelengths off a target and then measures the returned energy; for more information, see the sidebar "Remote Sensing" on page 30).

By bouncing radiation with wavelengths from 0.8 to 2.4 inches (2 to 6 cm) off Uranus, large-scale changes in the

This montage of images shows the planets visited by Voyager 2. (NASA/ JPL)

troposphere over a pressure range from about five to 50 bars can be imaged. Using these techniques, it has been shown that temperature gradients on Uranus change with latitude. The south pole appears to be hottest, and a distinct cooling in the troposphere occurs at a latitude of about −45 degrees.

In 1999 the *Hubble Space Telescope* took near-infrared images of Uranus that show giant cloud systems 1,250 miles (2,000 km) in diameter. These clouds appear to circle Uranus at speeds above 310 miles per hour (500 km/hr). The *Hubble*

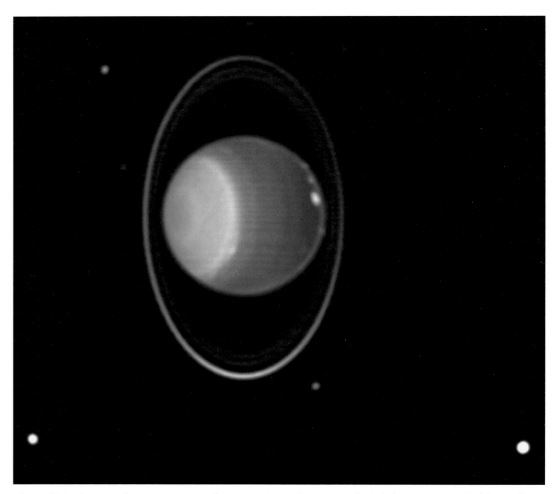

The Hubble Space Telescope *image of Uranus shown here reveals the planet surrounded by its four major rings and by 10 of its satellites.* (Kenneth Seidelmann, U.S. Naval Observatory, and NASA)

Space Telescope has seen as many as 20 bright clouds at different levels in the atmosphere. Most of these are visible only in the near infrared, but some can be seen in radio and visible wavelengths, as well. This image of Uranus demonstrates the large number of visible clouds and also shows 10 of Uranus's moons. Recently images from the Keck II telescope (completed in 1996 on Mauna Kea in Hawaii) have also shown large cloud features in the Uranian troposphere above the methane clouds. Uranus was long thought to have only mild winds, but the better resolution of the *Hubble Space Telescope* and of the Keck II telescope show that Uranus has immense winds, some as fast as 360 miles per hour (580 km/hr).

The clouds on Uranus are thought to consist mainly of methane because the other constituents of the atmosphere, hydrogen and helium, will not condense into fluid droplets or freeze into crystals at the temperature and pressure conditions of the Uranian atmosphere. The other trace molecules thought to contribute to cloud formation are ammonia (NH_3), ammonium hydrosulfide (NH_4SH), and hydrogen sulfide (H_2S). The location of the clouds is predicted based upon the temperature at which methane vapor will condense. Methane ice clouds are expected to form at pressures less than about one bar. Between about five bars and one bar, ammonia and hydrogen sulfide ice clouds should form. At pressures greater than five bars, clouds of ammonium hydrosulfide, water, ammonia, and hydrogen sulfide form, both alone and in solution with one another.

Both the *Hubble Space Telescope* and the Keck Observatory were recently able to resolve about 20 clouds on Uranus, nearly as many clouds on Uranus as the previous total in the history of modern observations. The cloud features seen in Keck and *Hubble* images are thousands of kilometers in diameter and exist at pressures from one to one-half bars. Improved imaging technology may be partly responsible, but more likely, cloud patterns on Uranus are highly dependent upon seasons, and the planet is moving into its second since humans have had the technology to see the planet.

Beyond the requirement of the most modern, high-resolution imaging techniques to see details in Uranian clouds, a

(continues on page 36)

REMOTE SENSING

Remote sensing is the name given to a wide variety of techniques that allow observers to make measurements of a place they are physically far from. The most familiar type of remote sensing is the photograph taken by spacecraft or by giant telescopes on Earth. These photos can tell scientists a lot about a planet; by looking at surface topography and coloration photo geologists can locate faults, craters, lava flows, chasms, and other features that indicate the weather, volcanism, and tectonics of the body being studied. There are, however, critical questions about planets and moons that cannot be answered with visible-light photographs, such as the composition and temperature of the surface or atmosphere. Some planets, such as Venus, have clouds covering their faces, and so even photography of the surface is impossible.

For remote sensing of solar system objects, each wavelength of radiation can yield different information. Scientists frequently find it necessary to send detectors into space rather than making measurements from Earth, first because not all types of electromagnetic radiation can pass through the Earth's atmosphere (see figure, opposite page), and second, because some electromagnetic emissions must be measured close to their sources, because they are weak, or in order to make detailed maps of the surface being measured.

Spectrometers are instruments that spread light out into spectra, in which the energy being emitted at each wavelength is measured separately. The spectrum often ends up looking like a bar graph, in which the height of each bar shows how strongly that wavelength is present in the light. These bars are called spectral lines. Each type of atom can only absorb or emit light at certain wavelengths, so the location and spacing of the spectral lines indicate which atoms are present in the object absorbing and emitting the light. In this way, scientists can determine the composition of something simply from the light shining from it.

Below are examples of the uses of a number of types of electromagnetic radiation in remote sensing.

GAMMA RAYS

Gamma rays are a form of electromagnetic radiation; they have the shortest wavelength and highest energy. High-energy radiation such as X-rays and gamma rays are absorbed to a great degree by the Earth's atmosphere, so it is not possible to measure their production by solar system bodies without sending measuring devices into space. These

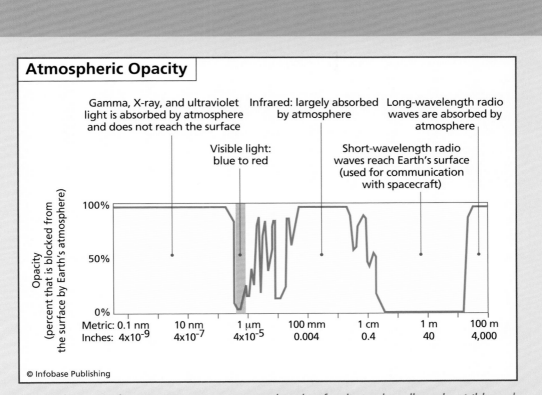

Atmospheric Opacity

Gamma, X-ray, and ultraviolet light is absorbed by atmosphere and does not reach the surface

Infrared: largely absorbed by atmosphere

Long-wavelength radio waves are absorbed by atmosphere

Visible light: blue to red

Short-wavelength radio waves reach Earth's surface (used for communication with spacecraft)

Opacity (percent that is blocked from the surface by Earth's atmosphere)

100%

50%

0%

Metric:	0.1 nm	10 nm	1 μm	100 mm	1 cm	1 m	100 m
Inches:	4×10^{-9}	4×10^{-7}	4×10^{-5}	0.004	0.4	40	4,000

© Infobase Publishing

The Earth's atmosphere is opaque to many wavelengths of radiation but allows the visible and short radio wavelengths through to the surface.

high-energy radiations are created only by high-energy events, such as matter heated to millions of degrees, high-speed collisions, or cosmic explosions. These wavelengths, then, are used to investigate the hottest regions of the Sun. The effects of gamma rays on other solar systems bodies, those without protective atmospheres, can be measured and used to infer compositions. This technique searches for radioactivity induced by the gamma rays.

Though in the solar system gamma rays are produced mainly by the hottest regions of the Sun, they can also be produced by colder bodies through a chain reaction of events, starting with high-energy cosmic rays. Space objects are continuously bombarded with cosmic rays, mostly high-energy protons. These high-energy protons strike the surface materials, such as dust and rocks, causing nuclear reactions in the atoms of the surface

(continues)

(continued)

material. The reactions produce neutrons, which collide with surrounding nuclei. The nuclei become excited by the added energy of neutron impacts, and reemit gamma rays as they return to their original, lower-energy state. The energy of the resultant gamma rays is characteristic of specific nuclear interactions in the surface, so measuring their intensity and wavelength allow a measurement of the abundance of several elements. One of these is hydrogen, which has a prominent gamma-ray emission at 2.223 million electron volts (a measure of the energy of the gamma ray). This can be measured from orbit, as it has been in the Mars Odyssey mission using a Gamma-Ray Spectrometer. The neutrons produced by the cosmic ray interactions discussed earlier start out with high energies, so they are called fast neutrons. As they interact with the nuclei of other atoms, the neutrons begin to slow down, reaching an intermediate range called epithermal neutrons. The slowing-down process is not too efficient because the neutrons bounce off large nuclei without losing much energy (hence speed). However, when neutrons interact with hydrogen nuclei, which are about the same mass as neutrons, they lose considerable energy, becoming thermal, or slow, neutrons. (The thermal neutrons can be captured by other atomic nuclei, which then can emit additional gamma rays.) The more hydrogen there is in the surface, the more thermal neutrons relative to epithermal neutrons. Many neutrons escape from the surface, flying up into space where they can be detected by the neutron detector on Mars Odyssey. The same technique was used to identify hydrogen enrichments, interpreted as water ice, in the polar regions of the Moon.

X-RAYS

When an X-ray strikes an atom, its energy can be transferred to the electrons orbiting the atom. This addition of energy to the electrons makes one or more electrons leap from their normal orbital shells around the nucleus of the atom to higher orbital shells, leaving vacant shells at lower energy values. Having vacant, lower-energy orbital shells is an unstable state for an atom, and so in a short period of time the electrons fall back into their original orbital shells, and in the process emit another X-ray. This X-ray has energy equivalent to the difference in energies between the higher and lower orbital shells that the electron moved between. Because each element has a unique set of energy levels between electron orbitals, each element produces X-rays with energies that are characteristic of itself and no other element. This method can be used remotely from a satellite, and it can also be used directly on tiny samples of material placed in a laboratory instrument called an electron microprobe, which measures the composition of the material based on the X-rays the atoms emit when struck with electrons.

VISIBLE AND NEAR-INFRARED

The most commonly seen type of remote sensing is, of course, visible light photography. Even visible light, when measured and analyzed according to wavelength and intensity, can be used to learn more about the body reflecting it.

Visible and near-infrared reflectance spectroscopy can help identify minerals that are crystals made of many elements, while other types of spectrometry identify individual types of atoms. When light shines on a mineral, some wavelengths are absorbed by the mineral, while other wavelengths are reflected back or transmitted through the mineral. This is why things have color to the eye: Eyes see and brains decode the wavelengths, or colors, that are not absorbed. The wavelengths of light that are absorbed are effectively a fingerprint of each mineral, so an analysis of absorbed versus reflected light can be used to identify minerals. This is not commonly used in laboratories to identify minerals, but it is used in remote sensing observations of planets.

The primary association of infrared radiation is heat, also called thermal radiation. Any material made of atoms and molecules at a temperature above absolute zero produces infrared radiation, which is produced by the motion of its atoms and molecules. At absolute zero, $-459.67°F$ ($-273.15°C$), all atomic and molecular motion ceases. The higher the temperature, the more they move, and the more infrared radiation they produce. Therefore, even extremely cold objects, like the surface of Pluto, emit infrared radiation. Hot objects, like metal heated by a welder's torch, emit radiation in the visible spectrum as well as in the infrared.

In 1879 Josef Stefan, an Austrian scientist, deduced the relation between temperature and infrared emissions from empirical measurements. In 1884 his student, Ludwig Boltzmann derived the same law from thermodynamic theory. The relation gives the total energy emitted by an object (E) in terms of its absolute temperature in Kelvin (T), and a constant called the Stefan-Boltzmann constant (equal to 5.670400×10^{-8} W m^{-2} K^{-4}, and denoted with the Greek letter sigma, σ):

$$E = \sigma T^4.$$

This total energy E is spread out at various wavelengths of radiation, but the energy peaks at a wavelength characteristic of the temperature of the body emitting the energy. The relation between wavelength and total energy, Planck's Law, allows scientists to determine the temperature of a body by measuring the energy it emits. The hotter the

(continues)

(continued)

body, the more energy it emits at shorter wavelengths. The surface temperature of the Sun is 9,900°F (5,500°C), and its Planck curve peaks in the visible wavelength range. For bodies cooler than the Sun, the peak of the Planck curve shifts to longer wavelengths, until a temperature is reached such that very little radiant energy is emitted in the visible range.

Humans radiate most strongly at an infrared wavelength of 10 microns (*micron* is another word for micrometer, one millionth of a meter). This infrared radiation is what makes night vision goggles possible: Humans are usually at a different temperature than their surroundings, and so their shapes can be seen in the infrared.

Only a few narrow bands of infrared light make it through the Earth's atmosphere without being absorbed, and can be measured by devices on Earth. To measure infrared emissions, the detectors themselves must be cooled to very low temperatures, or their own infrared emissions will swamp those they are trying to measure from elsewhere.

In thermal emission spectroscopy, a technique for remote sensing, the detector takes photos using infrared wavelengths and records how much of the light at each wavelength the material reflects from its surface. This technique can identify minerals and also estimate some physical properties, such as grain size. Minerals at temperatures above absolute zero emit radiation in the infrared, with characteristic peaks and valleys on plots of emission intensity versus wavelength. Though overall emission intensity is determined by temperature, the relationships between wavelength and emission intensity are determined by composition. The imager for *Mars Pathfinder*, a camera of this type, went to Mars in July 1997 to take measurements of light reflecting off the surfaces of Martian rocks (called reflectance spectra), and this data was used to infer what minerals the rocks contain.

When imaging in the optical or near-infrared wavelengths, the image gains information about only the upper microns of the surface. The thermal infrared gives information about the upper few centimeters, but to get information about deeper materials, even longer wavelengths must be used.

RADIO WAVES

Radio waves from outside the Earth do reach through the atmosphere and can be detected both day and night, cloudy or clear, from Earth-based observatories using huge metal dishes. In this way, astronomers observe the universe as it appears in radio waves. Images like photographs can be made from any wavelength of radiation coming from a body: Bright regions on the image can correspond to more intense radiation, and dark parts, to less intense regions. It is as if observers are looking at the object through eyes that "see" in the radio, or ultraviolet, or any other wavelength, rather than just visible. Because of a lingering feeling that humankind still observes the universe exclusively through our own eyes and ears, scientists still often refer to "seeing" a body in visible wavelengths and to "listening" to it in radio wavelengths.

Radio waves can also be used to examine planets' surfaces, using the technique called radar (radio detection and ranging). Radar measures the strength and round-trip

(continues)

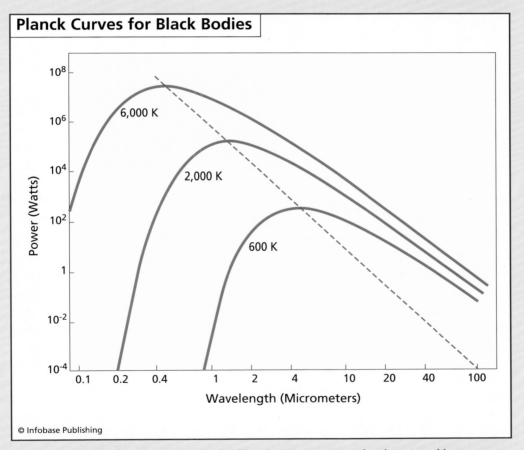

Planck Curves for Black Bodies

© Infobase Publishing

The infrared radiation emitted by a body allows its temperature to be determined by remote sensing; the curves showing the relationship between infrared and temperature are known as Planck curves.

(continued)

time of microwave or radio waves that are emitted by a radar antenna and bounced off a distant surface or object, thereby gaining information about the material of the target. The radar antenna alternately transmits and receives pulses at particular wavelengths (in the range 1 cm to 1 m) and polarizations (waves polarized in a single vertical or horizontal plane). For an imaging radar system, about 1,500 high-power pulses per second are transmitted toward the target or imaging area. At the Earth's surface, the energy in the radar pulse is scattered in all directions, with some reflected back toward the antenna. This backscatter returns to the radar as a weaker radar echo and is received by the antenna in a specific polarization (horizontal or vertical, not necessarily the same as the transmitted pulse). Given that the radar pulse travels at the speed of light, the measured time for the round trip of a particular pulse can be used to calculate the distance to the target.

Radar can be used to examine the composition, size, shape, and surface roughness of the target. The antenna measures the ratio of horizontally polarized radio waves sent to the surface to the horizontally polarized waves reflected back, and the same for vertically polarized waves. The difference between these ratios helps to measure the roughness of the surface. The composition of the target helps determine the amount of energy that is returned to the antenna: Ice is "low loss" to radar, in other words, the radio waves pass straight through it the way light passes through window glass. Water, on the other hand, is reflective. Therefore, by measuring the intensity of the returned signal and its polarization, information about the composition and roughness of the surface can be obtained. Radar can even penetrate surfaces and give information about material deeper in the target: By using wavelengths of 3, 12.6, and 70 centimeters, scientists can examine the Moon's surface to a depth of 32 feet (10 m), at a resolution of 330 to 985 feet (100 to 300 m), from the Earth-based U.S. National Astronomy and Ionosphere Center's Arecibo Observatory!

(continued from page 29)

second barrier to understanding Uranus's weather has simply been the length of its year, more than 84 Earth years. Although Uranus has been observed since its discovery more than 200 years ago, no one has ever seen this view of the planet in the modern era of astronomy. Because Uranus is tilted completely onto its side and orbits the Sun once every

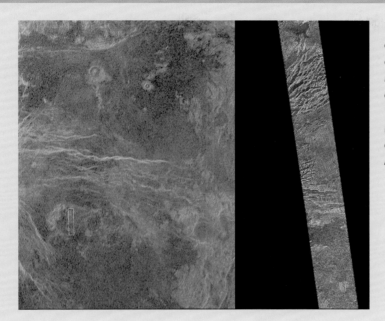

The far greater resolution obtained by the Magellan craft (right) shows the relative disadvantage of taking images of Venus from the Earth (left) using the Arecibo observatory. (NASA/ Magellan/JPL)

Venus is imaged almost exclusively in radar because of its dense, complete, permanent cloud cover. Radar images of Venus have been taken by several spacecraft and can also be taken from Arecibo Observatory on Earth. The image below makes a comparison between the resolution possible from Earth using Arecibo (left), and the resolution from the *Magellan* spacecraft (right). Arecibo's image is 560 miles (900 km) across and has a resolution of 1.9 miles (3 km). The *Magellan* image corresponds to the small white rectangle in the Arecibo image, 12 × 94 miles (20 × 120 km) in area. *Magellan*'s resolution is a mere 400 feet (120 m) per pixel.

84 years, it has 20-year-long seasons. The northern hemisphere of Uranus is at the end of its decades-long winter.

Early visual observers reported Jupiter-like cloud belts on the planet, but when NASA's *Voyager 2* flew by in 1986, Uranus appeared as featureless as a cue ball. This may have been due to where Uranus was in its year, locked in northern-hemisphere winter. Since then, the planet has moved far enough along its orbit for the Sun to shine at mid-latitudes in its northern hemisphere. Observations over a period of five years have shown changes in the temperature scales and patterns, and this initial data will allow better modeling of the Uranian atmosphere. It is apparent from these images that seasonal changes caused by Uranus's orbit do change wind and cloud patterns on the planet.

Rings and Moons
of Uranus

As with the other gas giant planets, many of the discoveries of rings and moon have occurred recently. Two of Uranus's moons were discovered more than 200 years ago, but as recently as 1986 only five moons had been discovered. Uranus's rings were discovered in 1977, the first ring system to be discovered during the modern age. Now Uranus is known to have 27 moons (more are likely to be found in coming years) and a set of narrow planetary rings.

RINGS

In 1977 Jim Elliot, a professor at the Massachusetts Institute of Technology, along with Robert Millis and Edward Dunham of the Lowell Observatory, watched a star move behind Uranus and in that moment discovered the planet's rings. In this technique, called stellar *occultation,* the quality of the starlight can help identify characteristics of the planet it is passing behind. The observations of Uranus were made using a 36-inch (91-cm) telescope aboard a Lockheed jet, together called the Gerard P. Kuiper Airborne Observatory. The jet flew at about 39,000 feet (12 km) to be above almost all of the Earth's atmospheric water vapor and provide clear measurements of infrared radiation during the occultation. In this case, the star's light blinked as it passed behind Uranus's rings, which were otherwise too small to be seen. Stellar occultation is also an effective technique for measuring a planet's atmosphere. Nine narrow rings were counted using stellar occultation, and then in 1986 *Voyager 2* detected two more: a narrow ring and a broad, diffuse ring closer to the planet.

Uranus has in total 11 rings, all in the planet's equatorial plane, that is, perpendicular to its orbit. They vary from ca. 6 to 60 miles (10 to 100 km) in width. The rings are designated using Greek letters, starting from the ring closest to Uranus:

> 1986 U2R
> 6
> 5
> 4
> α (alpha)
> β (beta)
> η (eta)
> γ (gamma)
> δ (delta)
> λ (lambda)
> ε (epsilon).

All the rings are listed and described in the table on page 41. The 10 outer rings are dark, thin, and narrow, and the 11th is broad and diffuse, inside the others. The ring 1986 U2R is the broad, diffuse ring. The densest and widest ring is the ε (epsilon) ring, with a width that varies from 12 to 60 miles (20 to 96 km). The other rings are just 0.6 to six miles (1 to 10 km) wide. All the rings are thin, less than 100 feet (a few tens of meters) in thickness perpendicular to the plane of their orbits.

Some scientists think that the dark color of the rings means they are old, perhaps because during aging the rings have lost their water, methane, or ammonia frost. By using occultation profiles at different wavelengths, the sizes of the particles in the rings can actually be determined. Uranus's rings consist mainly of particles between one and several tens of centimeters in diameter, much larger than the particles in Saturn's rings.

URANIAN RINGS			
Name	Distance from Uranus's surface to inner edge of ring (miles [km])	Width (miles [km])	Eccentricity
1986U2R	23,613 (38,000)	1,550 (2,500)	unavailable
6	25,997 (41,837)	~1 (~1.5)	0.0010
5	26,244 (42,235)	~1.2 (~2)	0.0019
4	26,453 (42,571)	~1.6 (~2.5)	0.0010
α (alpha)	27,787 (44,718)	2.5 to 6.2 (4 to 10)	0.0008
β (beta)	28,373 45,661)	3 to 7 (5 to 11)	0.0004
η (eta)	29,315 (47,176)	1 (1.6)	unavailable
γ (gamma)	29,594 (47,626)	0.6 to 2.5 (1 to 4)	0.0001
δ (delta)	30,015 (48,303)	1.9 to 4.3 (3 to 7)	unavailable
λ (lambda)	31,084 (50,024)	~1 (~2)	unavailable
ε (epsilon)	31,783 (51,149)	12 to 60 (20 to 96)	0.0079

Prior to 1977, only Saturn's broad, bright rings were known, and no astronomer seemed to have been thinking about the possibility of narrow, dark rings. How narrow rings are maintained is still not well understood. At first the theory was that small moons, called shepherd moons, orbit near the rings, confining the rings to narrow regions through gravitational pull from these moons. Uranus's moons Cordelia and Ophelia do indeed shepherd the ε ring. Other rings, though, seem to have no shepherds, and so the theories for how rings remain narrow are not completely delineated.

Strangely, the rings are not perfectly round. They have eccentricities from 0.001 to 0.01. This was another new phenomenon discovered in the Uranian rings: Previously, all physical theories for ring formation required that the rings be

(continues on page 44)

WHY ARE THERE RINGS?

Galileo Galilei first saw Saturn's rings in 1610, though he thought of them as handles on the sides of the planet rather than planet-encircling rings. After this first half observation, there was a hiatus in the discovery of planetary ring systems that lasted for three and a half centuries, until 1977, when Jim Elliot, an astronomer at the Massachusetts Institute of Technology and the Lowell Observatory, saw the blinking of a star's light as Uranus passed in front of it and correctly theorized that Uranus had rings around it that were blocking the light of the star. Two years later, *Voyager 1* took pictures of Jupiter's rings, and then in 1984, Earth-based observations found partial rings around Neptune. Now it is even hypothesized that Mars may have very tenuous rings, with an optical depth of more than 10^{-8} (meaning that almost all the light that shines on the ring goes straight through, without being scattered or reflected).

There are two basic categories of planetary rings. The first involves rings that are dense enough that only a small percentage of the light that shines on them passes through. These dense rings are made of large particles with radii that range from centimeters to meters. Examples of these dense rings are Saturn's main rings A and B and the Uranian rings. The second involves tenuous rings of particles the size of fine dust, just microns across. In these faint rings the particles are far apart, and almost all the light that strikes the ring passes through. Jupiter and Saturn's outermost rings are of this faint type. Neptune's rings, however, do not fall into either neat category.

In dense rings the constant collisions between particles act to spread out the rings. Particles near the planet lose speed when they collide with other particles, and thus fall closer to the planet. Particles at the outer edges of the ring tend to gain speed when they collide, and so move farther from the planet. In a complex way, the changes in velocity and redistributions of angular momentum act to make the ring thinner and thinner in depth while becoming more and more broadly spread from the planet.

Most dense rings exist within a certain distance from their planet, a distance called the Roche limit. Within the Roche limit, the tidal stresses from the planet's gravity overcome the tendency for particles to accrete into bodies: The gravitational stresses are stronger than the object's self-gravity, and the object is pulled to pieces. The Roche limit differs for every planet, since their gravities vary, and it also differs for each orbiting object, since their densities differ. The Roche limit (R_L) can be calculated as follows:

$$R_L = 2.446 R \left(\frac{\rho_{planet} e}{\rho_{satellite}} \right)^{\frac{1}{3}},$$

where ρ_{planet} is the density of the planet, $\rho_{satellite}$ is the density of the object orbiting the planet, and R is the radius of the planet. Thus, moons that attempted to form within the Roche limit, or were thrown within the Roche limit by other forces, will be torn into rubble by the gravitational forces of the planet and form rings.

All the Uranian rings lie within the Roche limit, but Saturn's, Jupiter's, and Neptune's outer rings lie outside their Roche limits and orbit in the same regions as moons. The moons have important effects on the rings near which they orbit. First, if the moon and particles in the ring share an orbital resonance (when the ratio of the orbital times is an integer ratio, for example, the moon orbits once for every two times the particle orbits), they interact gravitationally in a strong way: The particle, moon, and planet line up regularly, exerting strong forces on the particle and warping its orbit. If the resonances are strong (low integer ratios), a gap can be created in the ring. In other cases the resonance results in a wavy ring.

Moons can also strongly affect ring particles that they orbit next to. Very thin rings that would otherwise be expected to widen with time can be kept thin by moons that orbit closely on either side. These are called shepherd moons, bumping any stray particles back into the rings or accreting them onto the moon's surface.

Intact moons outside the Roche limit may also shed material, forming the source of a ring. Small moons, with low gravity, may allow more material to escape than large moons do. Jupiter's small moons Adrastea, Metis, Amalthea, and Thebe are all thought to create their own rings. In the Saturnian system, by contrast, the large, 250-km radius moon Enceladus is thought to be the creator and sustainer of Saturn's E ring.

Once formed, the ring does not remain forever: Forces from radiation, meteoroid impacts, and drag from the outer parts of the planet's atmosphere (the exosphere) begin to erode the ring. It is estimated that even dense rings can only exist for a few hundreds of thousands or millions of years, and so even the gorgeous rings of Saturn are probably just a fleeting phenomenon in the age of the solar system. Faint rings may disappear within thousands of years, unless replenished by a moon.

A system of rings around a planet can be thought of as a miniature reenactment of the original solar nebula: The planet is the giant mass at the center of a rotating system, much as the early Sun was, and the rings are the material rotating around it. Material is taken from moons to make rings, and other material is swept up by moons, a sort of recycling between moons and rings. This may be the reason that the outer planets have rings and the inner planets do not; the outer planets have great inventories of moons to create and sweep up rings, while the inner planets do not have enough moons.

(continued from page 41)

almost perfectly circular and that they lie in the equatorial plane of the planet. In addition to their eccentricities, Uranus's rings have inclinations to the planet's equatorial plane that range as high as 0.062 degrees.

MOONS

Before 1986, five of Uranus's moons had been observed. William Herschel, the discoverer of Uranus, found Oberon and Titania in 1787. Though Herschel searched for 40 years, he never found another moon of Uranus. In fact, his sightings of Oberon and Titania were not even confirmed by another astronomer for 10 years. In 1851 William Lassell, a British astronomer who started his professional life as a brewer, found Umbriel and Ariel and reported them to the Royal Astronomical Society. Another century later, in 1948, Gerald Peter Kuiper, the Dutch-born American astronomer (see more about the Kuiper belt later in this volume), found the moon Miranda by using the large, 81-inch (200-cm) reflecting telescope in the McDonald Observatory in Texas.

In 1986 the *Voyager 2* mission found 10 more moons (Cordelia, Ophelia, Bianca, Cressida, Desdemona, Juliet, Portia, Rosalind, Belinda, and Puck), all closer to the planet than the original five, and since 1986, five more have been seen from Earth-based observatories (see table on page 45). The first two of these, Caliban and Sycorax, found in 1987, are more than 10 times as far from Uranus as any of the previously known moons. In 1999 the moon 1986 U10 was found sharing an orbit almost identical to that of Belinda. (Though this moon was found in 1999, it was then recognized on images from *Voyager* taken in 1986, hence its designation.) About once a month, Belinda laps 1986 U10 as they orbit! Since then, a moon (2003 U2) has been found with an orbit even closer to Belinda's than is 1986 U10. Recently discovered moons still have their year and number designations, and have not yet been given their permanent names. Uranus's moons are all named for characters from the writings of William Shakespeare and Alexander Pope. Uranus has, at the

URANIAN MOONS						
Moon	Orbital period [Earth days]	Radius [miles, (km)]	Year discovered	Orbital eccentricity	Orbital inclination [°]	Orbital direction
Regular Group						
1. Cordelia	0.335	13 (20)	1986	0.00026	0.085	prograde
2. Ophelia	0.376	14 (21)	1986	0.00992	0.104	prograde
3. Bianca	0.435	16 (25)	1986	0.00092	0.19	prograde
4. Cressida	0.464	25 (40)	1986	0.00036	0.005	prograde
5. Desdemona	0.474	20 (32)	1986	0.00013	0.111	prograde
6. Juliet	0.493	29 (47)	1986	0.00066	0.065	prograde
7. Portia	0.513	42 (68)	1986	0.00005	0.059	prograde
8. Rosalind	0.558	23 (36)	1980	0.00011	0.279	prograde
9. Cupid (2003 U2)	0.618	5 (9)	2003	0.0013	0.1	prograde
10. Belinda	0.624	28 (45)	1986	0.00007	0.031	prograde
11. Perdita (1986 U10)	0.638	9 (15)	1986	0.0012	0.00	prograde
12. Puck	0.762	50 (81)	1985	0.00012	0.319	prograde
13. Mab (2003 U1)	0.923	8 (13)	2003	0.0025	0.134	prograde
14. Miranda	1.414	146 (236)	1948	0.0013	4.23	prograde
15. Ariel	2.520	360 (579)	1851	0.0012	0.26	prograde
16. Umbriel	4.144	363 (584)	1851	0.0050	0.21	prograde
17. Titania	8.706	493 (789)	1787	0.0011	0.34	prograde
18. Oberon	13.46	476 (761)	1787	0.0014	0.06	prograde

(continues)

URANIAN MOONS (continued)

Moon	Orbital period [Earth days]	Radius [miles, (km)]	Year discovered	Orbital eccentricity	Orbital inclination [°]	Orbital direction
Irregular Group						
19. Francisco (2001 U3)	266.6	7 (11)	2001	0.146	147.5	retrograde
20. Caliban	579.5	23 (36)	1997	0.159	139.9	retrograde
21. Stephano	677.4	10 (16)	1999	0.229	141.9	retrograde
22. Trinculo	759.0	6 (9)	2001	0.220	166.25	retrograde
23. Sycorax	1288.3	47 (75)	1997	0.522	152.46	retrograde
24. Margaret (2003 U3)	1,694.8	6 (10)	2003	0.661	51.46	prograde
25. Prospero	1,977.3	19 (30)	1999	0.445	146.0	retrograde
26. Setebos	2,234.8	19 (30)	1999	0.591	145.88	retrograde
27. Ferdinand (2001 U2)	2,823.4	7 (12)	2001	0.426	167.34	retrograde

moment, 27 known moons, three of which were discovered in 2003. More moons will likely be discovered in the future. Most of the moons are almost perfectly in the plane of Uranus's equator, each within 0.4 degrees of the equatorial plane. Observing the orbits of the moons was the first indication of Uranus's highly tilted axis (Uranus's rotation axis lies 97.92 degrees away from perpendicular to its orbital plane).

Although little is known about many of the Uranian moons, the following is a summary of each moon, numbered in order of its orbit, starting closest to Uranus. When a moon has been given a provisional designation, and then a final name, its provisional designation is shown in parentheses. It is important to keep in mind that new satellites are being discovered at a high rate, and that new information on the sizes and orbits of existing satellites is made available frequently, and so when absolute up-to-date information is needed, one should consult the NASA, Hawaii Observatory, Minor Planet Center, or other Web sites.

Many of Uranus's moons are described here in more detail, but very little is known about some of the smaller moons. A few of the small, irregular moons have virtually nothing known about them and so are left out of this section.

1. Cordelia [1986 U7]
Cordelia is named after one of Lear's daughters in Shakespeare's *King Lear*. The moon was discovered by *Voyager 2* in 1986. Cordelia appears to be the inner shepherding moon for Uranus's ε ring, keeping the ring defined on its inner edge. Cordelia orbits inside the *synchronous orbit radius* for Uranus, so it orbits Uranus more than once in a Uranian day. Like many moons, Cordelia rotates synchronously, so the same face of Cordelia is always toward Uranus (see the sidebar "What Are Synchronous Orbits and Synchronous Rotation?" on page 54). Cordelia, and all the satellites through Rosalind, were discovered by *Voyager 2* in 1986.

2. Ophelia [1986 U8]
Ophelia is named after the daughter of Polonius in Shakespeare's *Hamlet*. Like Cordelia, the moon was discovered by *Voyager 2* in 1986, and it too orbits Uranus more than once per Uranian day. It appears to be the outer shepherding moon for Uranus's ε ring, keeping the ring defined at its outer edge.

3. Bianca [1986 U9]
Bianca is named for the sister of Katherine in Shakespeare's *The Taming of the Shrew*. Like many other Uranian moons, its density and mass are unknown, so its composition cannot be accurately guessed.

4. Cressida [1986 U3]
Cressida is named for the daughter of Calchas in Shakespeare's *Troilus and Cressida*. All the moons through Cressida orbit closer to Uranus than the Roche limit. Will they be torn apart

WHAT ARE SYNCHRONOUS ORBITS AND SYNCHRONOUS ROTATION?

Synchronous rotation can easily be confused with *synchronous orbits*. In a synchronous orbit, the moon orbits always above the same point on the planet it is orbiting (this section uses the terms *moon* and *planet,* but the same principles apply to a planet and the Sun). There is only one orbital radius for each planet that produces a synchronous orbit. Synchronous rotation, on the other hand, is created by the period of the moon's rotation on its axis being the same as the period of the moon's orbit around its planet, and produces a situation where the same face of the moon is always toward its planet. *Tidal locking* causes synchronous rotation.

Gravitational attraction between the moon and its planet produces a tidal force on each of them, stretching each very slightly along the axis oriented toward its partner. In the case of spherical bodies, this causes them to become slightly egg-shaped; the extra stretch is called a tidal bulge. If either of the two bodies is rotating relative to the other, this tidal bulge is not stable. The rotation of the body will cause the long axis to move out of alignment with the other object, and the gravitational force will work to reshape the rotating body. Because of the relative rotation between the bodies, the tidal bulges move around the rotating body to stay in alignment with the gravitational force between the bodies. This is why ocean tides on Earth rise and fall with the rising and setting of its moon, and the same effect occurs to some extent on all rotating orbiting bodies.

over time? (For more on the Roche limit, see the sidebar "Why Are There Rings?" on page 42) These moons are probably small enough to avoid destruction through tidal forces within the Roche limit, but they are also too small for clear observation and discovery of their compositions or surface characteristics.

5, 6, 7, 8. Desdemona [1986 U6], Juliet [1986 U2], Portia [1986 U1], Rosalind [1986 U4]

As with the inner satellites, almost nothing is known about these four small bodies. Desdemona is named for the wife of Othello in Shakespeare's *Othello*. Juliet was, of course, named after the heroine in Shakespeare's *Romeo and Juliet*. Portia

The rotation of the tidal bulge out of alignment with the body that caused it results in a small but significant force acting to slow the relative rotation of the bodies. Since the bulge requires a small amount of time to shift position, the tidal bulge of the moon is always located slightly away from the nearest point to its planet in the direction of the moon's rotation. This bulge is pulled on by the planet's gravity, resulting in a slight force pulling the surface of the moon in the opposite direction of its rotation. The rotation of the satellite slowly decreases (and its orbital momentum simultaneously increases). This is in the case where the moon's rotational period is faster than its *orbital period* around its planet. If the opposite is true, tidal forces increase its rate of rotation and decrease its orbital momentum.

Almost all moons in the solar system are tidally locked with their primaries, since they orbit closely and tidal force strengthens rapidly with decreasing distance. In addition, Mercury is tidally locked with the Sun in a 3:2 *resonance*. Mercury is the only solar system body in a 3:2 resonance with the Sun. For every two times Mercury revolves around the Sun, it rotates on its own axis three times. More subtly, the planet Venus is tidally locked with the planet Earth, so that whenever the two are at their closest approach to each other in their orbits, Venus always has the same face toward Earth (the tidal forces involved in this lock are extremely small). In general any object that orbits another massive object closely for long periods is likely to be tidally locked to it.

was named after an heiress in Shakespeare's *The Merchant of Venice*. Rosalind was named after a daughter of the banished duke in Shakespeare's *As You Like It*.

9. Cupid [2003 U2]
Mark Showalter, a research scientist at Stanford University, and Jack Lissauer, a research scientist at the NASA Ames Research Center, discovered 2003 U1 and 2003 U2 with the *Hubble Space Telescope*.

10. Belinda [1986 U5]
Belinda is named after the heroine in Alexander Pope's *The Rape of the Lock* and was discovered by *Voyager 2* in 1986.

11. Perdita [1986 U10]

This tiny moon was imaged by *Voyager 2* in 1986, but not noticed in the images until 1999, by Erich Karkoschka of the Lunar and Planetary Lab of the University of Arizona in Tucson. This moon's orbit is nearly identical to that of Belinda, about 47,000 miles (75,000 km) from Uranus. For several years the moon's existence was in doubt because it was so small that it could not be seen with Earth-based telescopes. In fact, after initially being given its temporary designation and announced as a new moon of Uranus, the International Astronomical Union decided there was not enough data on the object, and stripped it of the title "moon." In 2003 it was finally seen again in an image made by the newest camera on the *Hubble Space Telescope,* its existence confirmed, and its status as a moon of Uranus renewed.

12. Puck [1985 U1]

Puck is a playful fairy in Shakespeare's *A Midsummer Night's Dream*. Of the 10 new Uranian moons discovered by *Voyager 2,* only Puck was discovered soon enough for the mission's observation schedule to be adjusted to get images. Puck is irregularly shaped, with a mottled surface, and it and the other small moons are very dark, each with an *albedo* less than 0.1.

13. Mab [2003 U1]

Mark Showalter and Jack Lissauer discovered 2003 U1 with the *Hubble Space Telescope*.

14. Miranda

Miranda is named after the daughter of the magician Prospero in Shakespeare's *The Tempest*. In 1948 Gerald Kuiper, the Dutch-born American astronomer, discovered the moon. Miranda was the last moon of Uranus discovered before *Voyager 2. Voyager 2* flew close to Uranus to get the boost it needed to go on to Neptune, and happened to pass close to Miranda, obtaining good images of the moon.

Miranda is the smallest of Uranus's major moons, but it has the most complex and interesting surface. At 290 miles (470 km) in radius, it seems to be at size interface between

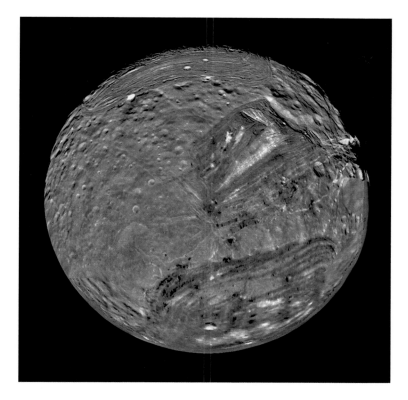

Uranus's moon Miranda displays its distinctive chevron feature above and to the right of center in this image centered on its south pole. (NASA/ JPL/USGS/Voyager)

planets that differentiate (and turn into proper spheres) and small, irregular planets. Miranda's surface is heavily cratered, as expected for an old surface on a small body, but the old surface is covered with unusual patterns. Miranda has lines and grooves (called "ovoids"), as well as a white chevron-shaped cracks, and three strange areas called coronae that have been geologically active. Miranda's image above shows both old, heavily cratered, rolling terrain and young, complex terrain containing bright and dark bands and ridges.

Each of these three coronae has clearly defined edges and lacks cratering, showing that they formed after the heavy period of cratering early in the solar system. The largest corona, Arden, is a giant oval of concentric light and dark stripes, bounded by a deep canyon. The second-largest corona, called Inverness, covers Miranda's south pole and is rectangular, though also covered with stripes and ridges and

(continues on page 56)

ACCRETION AND HEATING: WHY ARE SOME SOLAR SYSTEM OBJECTS ROUND AND OTHERS IRREGULAR?

There are three main characteristics of a body that determine whether it will become round.

The first is its *viscosity,* that is, its ability to flow. Fluid bodies can be round because of surface tension, no matter their size; self-gravitation does not play a role. The force bonding together the molecules on the outside of a fluid drop pull the surface into the smallest possible area, which is a sphere. This is also the case with gaseous planets, like Uranus. Solid material, like rock, can flow slowly if it is hot, so heat is an important aspect of viscosity. When planets are formed, it is thought that they start as agglomerations of small bodies, and that more and more small bodies collide or are attracted gravitationally, making the main body larger and larger. The heat contributed by colliding planetesimals significantly helps along the transformation of the original pile of rubble into a spherical planet: The loss of their kinetic energy (more on this at the end of this sidebar) acts to heat up the main body. The hotter the main body, the easier it is for the material to flow into a sphere in response to its growing gravitational field.

The second main characteristic is density. Solid round bodies obtain their shape from gravity, which acts equally in all directions and therefore works to make a body a sphere. The same volume of a very dense material will create a stronger gravitational field than a less dense material, and the stronger the gravity of the object, the more likely it is to pull itself into a sphere.

The third characteristic is mass, which is really another aspect of density. If the object is made of low-density material, there just has to be a lot more of it to make the gravitational field required to make it round.

Bodies that are too small to heat up enough to allow any flow, or to have a large enough internal gravitational field, may retain irregular outlines. Their shapes are determined by mechanical strength and response to outside forces such as meteorite impacts, rather than by their own self-gravity. In general the largest asteroids, including all 100 or so that have diameters greater than 60 miles (100 km), and the larger moons, are round from self-gravity. Most asteroids and moons with diameters larger than six miles (10 km) are round, but not all of them, depending on their composition and the manner of their creation.

There is another stage of planetary evolution after attainment of a spherical shape: internal differentiation. All asteroids and the terrestrial planets probably started out made of primitive materials, such as the class of asteroids and meteorites called CI or enstatite chondrites. The planets and some of the larger asteroids then became compositionally stratified in their interiors, a process called differentiation. In a *differentiated body,* heavy metals, mainly iron with some nickel and other minor impurities in the case of terrestrial planets, and rocky and icy material in the case of the gaseous planets, have sunk to the middle of the body, forming a core. Terrestrial planets are therefore made up, in a rough sense, of concentric shells of materials with different compositions. The outermost shell is a crust, made mainly of material that has melted from the interior and risen buoyantly up to the surface. The mantle is made of silicate minerals, and the core is mainly of iron. The gas giant outer planets are similarly made of shells of material, though they are gaseous materials on the outside and rocky or icy in the interior. Planets with systematic shells like these are called differentiated planets. Their concentric spherical layers differ in terms of composition, heat, density, and even motion, and planets that are differentiated are more or less spherical. All the planets in the solar system seem to be thoroughly differentiated internally, with the possible exception of Pluto and Charon. What data there is for these two bodies indicates that they may not be fully differentiated.

Some bodies in the solar system, though, are not differentiated; the material they are made of is still in a more primitive state, and the body may not be spherical. Undifferentiated bodies in the asteroid belt have their metal component still mixed through their silicate portions; it has not separated and flowed into the interior to form a core.

Among asteroids, the sizes of bodies that differentiated vary widely. Iron meteorites, thought to be the differentiated cores of rocky bodies that have since been shattered, consist of crystals that grow to different sizes directly depending upon their cooling rate, which in turn depends upon the size of the body that is cooling. Crystal sizes in iron meteorites indicate parent bodies from six to 30 miles (10 to 50 km) or more in diameter. Vesta, an asteroid with a basaltic crust and a diameter of 326 miles (525 km), seems to be the largest surviving differentiated body in the asteroid belt. Though the asteroid Ceres, an unevenly-shaped asteroid approximately 577 by 596 miles (930 by

(continues)

(continued)

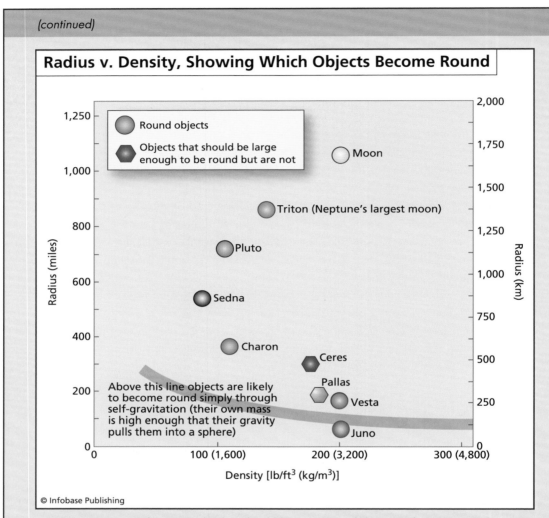

Radius v. Density, Showing Which Objects Become Round

Round objects

Objects that should be large enough to be round but are not

Moon

Triton (Neptune's largest moon)

Pluto

Sedna

Charon

Ceres

Pallas

Vesta

Juno

Above this line objects are likely to become round simply through self-gravitation (their own mass is high enough that their gravity pulls them into a sphere)

Radius (miles)

Radius (km)

Density [lb/ft³ (kg/m³)]

© Infobase Publishing

At a certain mass, solar system bodies should become round due to self-gravitation. This general rule works for some bodies, but others have remained irregular despite what should be a large enough mass.

960 km), is much larger than Vesta, it seems from spectroscopic analyses to be largely undifferentiated. It is thought that the higher percentages of volatiles available at the distance of Ceres's orbit may have helped cool the asteroid faster and prevented the buildup of heat required for differentiation. It is also believed that Ceres and Vesta are among the last surviving "protoplanets," and that almost all asteroids of smaller size are the shattered remains of larger bodies.

Where does the heat for differentiation come from? The larger asteroids generated enough internal heat from radioactive decay to melt (at least partially) and differentiate (for more on radioactive decay, see the sidebar called "Elements and Isotopes" on page 20). Generally bodies larger than about 300 miles (500 km) in diameter are needed in order to be insulated enough to trap the heat from radioactive decay so that melting can occur. If the body is too small, it cools too fast and no differentiation can take place.

A source for heat to create differentiation, and perhaps the main source, is the heat of accretion. When smaller bodies, often called *planetesimals,* are colliding and sticking together, creating a single larger body (perhaps a planet), they are said to be accreting. Eventually the larger body may even have enough gravity itself to begin altering the paths of passing planetesimals and attracting them to it. In any case, the process of accretion adds tremendous heat to the body, by the transformation of the kinetic energy of the planetesimals into heat in the larger body. To understand kinetic energy, start with momentum, called *p,* and defined as the product of a body's mass *m* and its velocity *v:*

$$p = mv$$

Sir Isaac Newton called momentum "quality of movement." The greater the mass of the object, the greater its momentum is, and likewise, the greater its velocity, the greater its momentum is. A change in momentum creates a force, such as a person feels when something bumps into her. The object that bumps into her experiences a change in momentum because it has suddenly slowed down, and she experiences it as a force. The reason she feels more force when someone tosses a full soda to her than when they toss an empty soda can to her is that the full can has a greater mass, and therefore momentum, than the empty can, and when it hits her it loses all its momentum, transferring to her a greater force.

How does this relate to heating by accretion? Those incoming planetesimals have momentum due to their mass and velocity, and when they crash into the larger body, their momentum is converted into energy, in this case, heat. The energy of the body, created by its mass and velocity, is called its kinetic energy. Kinetic energy is the total effect of changing momentum of a body, in this case, as its velocity slows down to zero. Kinetic energy is expressed in terms of mass *m* and velocity *v:*

$$K = \frac{1}{2}mv^2$$

(continues)

(continued)

Students of calculus might note that kinetic energy is the integral of momentum with respect to velocity:

$$K = \int mv\,dv = \frac{1}{2}mv^2$$

The kinetic energy is converted from mass and velocity into heat energy when it strikes the growing body. This energy, and therefore heat, is considerable, and if accretion occurs fast enough, the larger body can be heated all the way to melting by accretional kinetic energy. If the larger body is melted even partially, it will differentiate.

How is energy transfigured into heat, and how is heat transformed into melting? To transfer energy into heat, the type of material has to be taken into consideration. Heat capacity describes how a material's temperature changes in response to added energy. Some materials go up in temperature easily in response to energy, while others take more energy to get hotter. Silicate minerals have a heat capacity of 245.2 cal/°lb (1,256.1 J/°kg). What this means is that 245.2 calories of energy are required to raise the temperature of one pound of silicate material one degree. Here is a sample calculation. A planetesimal is about to impact a larger body, and the planetesimal is a kilometer in radius. It would weigh roughly 3.7×10^{13} lb (1.7×10^{13} kg), if its density were about 250 lb/ft³ (4,000 kg/m³). If it were traveling at six miles per second (10 km/s), then its kinetic energy would be

$$K = \frac{1}{2}mv^2 = \left(1.7 \times 10^{13}\,kg\right)\left(10,000\,m/sec\right)^2$$
$$= 8.5 \times 10^{20}\,J = 2 \times 10^{20}\,cal$$

Using the heat capacity, the temperature change created by an impact of this example planetesimal can be calculated:

(continued from page 51)

troughs of dark and light material. Its stripes are in the shape of a right angle, filling in the corona from one side. The third corona, Elsinore, is oval, like Arden, but filled with troughs and ridges, like Miranda, though it lacks the color striping and appears plain gray.

$$\frac{8.5 \times 10^{20°} kg}{1,256.1 J / °kg} = 6.8 \times 10^{17°} kg = 8.3 \times 10^{17°} lb$$

The question now becomes, how much mass is going to be heated by the impact? According to this calculation, the example planetesimal creates heat on impact sufficient to heat one pound of material by $8.3 \times 10^{17}°F$ (or one kilogram by $6.8 \times 10^{17}°C$), but of course it will actually heat more material by lesser amounts. To calculate how many degrees of heating could be done to a given mass, divide the results of the previous calculation by the mass to be heated.

The impact would, of course, heat a large region of the target body as well as the impactor itself. How widespread is the influence of this impact? How deeply does it heat, and how widely. Of course, the material closest to the impact will receive most of the energy, and the energy input will go down with distance from the impact, until finally the material is completely unheated. What is the pattern of energy dispersal? Energy dispersal is not well understood even by scientists who study impactors.

Here is a simpler question: If all the energy were put into melting the impacted material, how much could it melt? To melt a silicate completely requires that its temperature be raised to about 2,700°F (1,500°C), as a rough estimate, so here is the mass of material that can be completely melted by this example impact:

$$\frac{6.8 \times 10^{17°} kg}{1,500°} = 4.5 \times 10^{14°} kg = 9.9 \times 10^{14} lb$$

This means that the impactor can melt about 25 times its own mass ($4.5 \times 10^{14}/1.7 \times 10^{13} = 26$). Of course this is a rough calculation, but it does show how effective accretion can be in heating up a growing body, and how it can therefore help the body to attain a spherical shape and to internally differentiate into different compositional shells.

Canyons hundreds of kilometers long and tens of kilometers wide score other parts of Miranda's surface. These canyons appear to have been formed by stretching Miranda's crust, perhaps because buoyant upwelling material from deep inside Miranda rose to the bottom of the crust and lifted and stretched it. There are many theories about the formation of the coronae, as well. Some researchers propose that the coronae are sequential outpourings of lava from these hot upwellings. Others argue that the coronae look very much like oceanic crust on Earth and are the frozen remnants of *plate tectonics* on Miranda. *Voyager 2* acquired the image of Miranda on page 55 from a distance of 19,400 miles (31,000 km) but achieved a high resolution of about 2,000 feet (600 m) per pixel. The grooves reach depths of a few kilometers. The image area is about 150 miles (240 km) across.

All these theories have a fundamental mystery: Where did the heat come from to cause buoyant upwellings, volcanic activity, and plate tectonics? Miranda is too small to have built up any significant heat by accretion (see the sidebar "Accretion and Heating" on page 52), and all its heat would have been lost very quickly, also because of its small size. Miranda appears to have a density of only 70 lb/ft³ (1,115 kg/m³), indicating that it consists largely of water ice. Its crust may indeed have lava flows and plate tectonics, but the molten material would have to consist mostly of water. Unfortunately, the formations that are thought to be lava are hundreds of meters thick and have sharp ridges. Both these features indicate that the lava had to have a high viscosity (that is, be a thick and slowly flowing liquid), which liquid water would not be. If the lava consisted of water, it would be thin and run out into sheets, not able to build up into thick flows and ridges. It is possible, though, that the lava consisted of a mixture of water and ammonia or methanol, which at the very low surface temperatures of Miranda (around −310°F or −190°C) would have a reasonably high viscosity and would freeze like rock.

All these theories about Miranda are currently unsubstantiated and await more space missions to gather more data. Miranda's very strange surface, apparently so active in the past, may have been created by processes so unlike those on Earth that scientists have not yet imagined the right hypotheses.

15. Ariel

Ariel is named for an ethereal spirit in Shakespeare's *The Tempest* and was discovered by William Lassell, a British astrono-

Ariel's complex surface is thought to have been made by a combination of interior and surface processes along with cratering. (NASA/JPL/ Voyager 2)

mer who started his professional life as a brewer, in 1851. Ariel has the youngest surface of the five major moons, and is the brightest. Its appearance is similar to Titania's. It bears many deep, flat-floored canyons crossing back and forth over an older, cratered surface. The largest canyon is called Kachina Chasmata and is 390 miles (622 km) long. The floors of the canyons appear to have volcanic activity in them, and the surface also bears thin, winding, raised ridges called rilles, which on Earth's Moon have been identified as lava tubes formed during spreading volcanic activity.

The images for the montage of Ariel on page 58 were taken by *Voyager 2* on January 24, 1986, at a distance of about 80,000 miles (130,000 km). Ariel's many small craters are close to the threshold of detection in this picture. Numerous valleys and fault scarps crisscross the highly pitted terrain. The largest fault valleys, at the right, as well as a smooth region near the center of this image, have been partly filled with younger deposits that are less heavily cratered than the pitted terrain. Narrow, somewhat sinuous scarps and valleys have been formed, in turn, in these young deposits.

On Ariel, the volcanic activity is thought to be not the hot, silica-based molten rocks found on Earth, Io, and Mars but a type of cold volcanism *(cryovolcanism)* created by a flowing mixture of ammonia and water at very low temperatures. Though Ariel currently has no tidal resonance with other moons, it is thought that the cryovolcanism was caused in the past by tidal heating, and that the deep canyons formed when Ariel cooled from its initial formation. The interior is thought to be water-rich, so scientists hypothesize that when the liquid interior cooled and froze, it expanded (water is one of the very few materials that actually expands when it freezes), splitting Ariel's crust.

16. Umbriel

Umbriel, named after a character in Alexander Pope's *The Rape of the Lock,* was discovered by William Lassell in 1851. Umbriel is thought to be in its primitive state, never having heated enough to differentiate into a core and a mantle. Its surface is very dark (only half as bright as Ariel), heavily cratered, and appears never to have been resurfaced. Its dark, cratered surface appears similar to Oberon's and very different from Titania's and Ariel's. Its outstanding feature is the crater Wunda, 50 miles (80 km) in radius, which has a conspicuously bright floor.

17. Titania

Titania is named after the Queen of the Fairies, wife of Oberon, in Shakespeare's *A Midsummer Night's Dream.* The largest of Uranus's moons (see image on page 58), it was discovered by William Herschel in 1787. Titania is an icy moon with relatively few craters on its surface, indicating that it may have been resurfaced at some point in its history. Its appearance is similar to Ariel's. At 100 miles (163 km) in radius, Gertrude is the moon's largest crater. Large, interconnected canyons, seemingly caused by faulting, cover Titania's surface. The

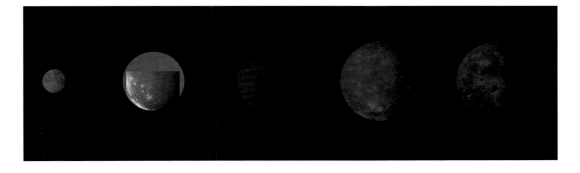

largest canyon is Messina Chasmata, 940 miles (1,500 km) long and 60 miles (100 km) across at its widest point.

18. Oberon

Oberon is named after the King of the Fairies, husband of Titania, in Shakespeare's *A Midsummer Night's Dream*. Discovered by William Herschel in 1787, Oberon is a midsize moon and one of the original five discovered before the *Voyager 2* mission. Its surface is heavily cratered with what may be volcanic flows on some large crater floors. Its dark, cratered surface is similar to Umbriel's. Some craters have light-colored *ejecta* rays radiating from them. Oberon's largest crater is Hamlet, at 60 miles (100 km) in radius. The images from *Voyager 2* were not as good as those for other moons, but scientists did make out a 310-mile (500-km)-long linear feature, Mommur Chasma, that appears to be created by a fault. The *Voyager 2* image shown on page 59, taken from distance of 1.72 million miles (2.77 million km), clearly displays large craters.

The surface of Oberon seems to be very old, not resurfaced with water-rich lavas as Miranda and Ariel seem to have been. Oberon's leading hemisphere appears redder than its other hemisphere (like Earth's Moon, its rotation is synchronous, that is, the same side of the moon faces Uranus at all times, so one hemisphere of Oberon always leads in its orbit). The reddish color may be due to dust from other moons swept up by Oberon during orbit.

All the moons with orbits outside Oberon are irregular: They have large orbits at significant inclinations, and almost all of these moons orbit in a retrograde sense.

Uranus's five largest satellites, Miranda, Ariel, Umbriel, Titania, and Oberon, are shown in this montage from left to right (also in order of increasing distance from Uranus). The moons are presented at the same size and brightness scales to allow comparisons. Photographic coverage is incomplete for Miranda and Ariel, and gray circles depict missing areas. (NASA/JPL/ Voyager 2)

Uranus's moon Oberon displays some bright patches suggestive of crater ejecta in an icy surface. (NASA/JPL/ Voyager 2)

19. Francisco [2001 U3]

Matthew Holman and J. J. Kavelaars first observed this tiny moon using the 13-foot (4-m) Blanco telescope at the Cerro Tololo Inter-American Observatory in Chile. Although it was first seen in 2001, it was not confirmed until 2003.

20. Caliban [1997 U1]

Caliban is named after a savage slave from Shakespeare's play *The Tempest,* the son of the witch Sycorax who imprisoned the fairy Ariel for disobedience. In 1997 Brett Gladman, a scientist at the Canadian Institute for Theoretical Astrophysics and Cornell University, and his colleagues Phil Nicholson, Joseph Burns, and J. J. Kavelaars discovered Caliban using the 200-inch (5-m) Hale telescope. The first images of the moon were taken on September 6 and 7 of that year. Prior to the discovery of Caliban and Sycorax, all of Uranus's moons orbited directly (in the same sense as the Earth orbits the Sun) and in planes close to the planet's equator.

All the other gas giant planets were known to have irregular moons with unusual orbits, and now so did Uranus: Both Caliban and Sycorax have highly inclined, retrograde orbits, and both are assumed to be captured asteroids. Both are reddish in color, similar to some Kuiper belt objects. At the time of their discovery, they were the dimmest moons ever to be discovered by a ground-based observatory; Caliban is slightly dimmer than Sycorax. Almost all the moons exterior to Caliban have retrograde orbits and are likely candidates to be captured asteroids. The moons of all the gas giants fall into this pattern: Near moons have more circular, less inclined, prograde orbits, and farther moons have inclined, eccentric orbits and are candidates to be captured asteroids.

21. Stephano [1999 U2]
Almost nothing is known about this small moon; it was found at the same time as Setebos.

22. Trinculo [2001 U1]
Trinculo is named for the jester in Shakespeare's play *The Tempest*. The productive team of Brett Gladman, Matthew Holman, J. J. Kavelaars, and Jean-Marc Petit found this tiny moon, dimmer than many of those that were found by *Voyager 2* when it was near Uranus.

23. Sycorax [1997 U2]
Sycorax is named after the witch in Shakespeare's play *The Tempest* who is also the mother of Caliban. In 1997 Phil Nicholson, Brett Gladman, Joseph Burns, and J. J. Kavelaars discovered Sycorax using the 200-inch (5-m) Hale telescope. See the entry on Caliban for more details.

24. Margaret [2003 U3]
Scott S. Sheppard and Dave Jewitt at the University of Hawaii discovered two new irregular moons of Uranus, 2001 U2 and 2003 U3, from images obtained by the Subaru 326-inch (8.3-m) telescope at Mauna Kea in Hawaii, on August 29, 2003. Additional observations by the Hawaii team using the Gemini 322-inch (8.2-m) telescope allowed Brian Marsden at the Minor Planet Center to correlate one of the moons to an independent discovery made in 2001 by a group led by Matthew Holman and J. J. Kavelaars. The 2001 observations were not sufficient on their own to determine if the objects were satellites of Uranus; no reliable orbits were found. They were then lost until discovery in 2003 by the Hawaii team. The

moon 2003 U3 has the first direct (prograde) orbit of any of Uranus's irregular satellites.

25. Prospero [1999 U3]
Almost nothing is known about this small moon; it was found at the same time as Setebos.

26, 27. Setebos [1999 U1] and Ferdinand [2001 U2]
Very little is known about these additional irregular moons.

As of the writing of this chapter, Uranus is known to have 27 moons. This number is almost certain to rise with time. Jupiter is known to have 63 moons, and Saturn is known to have 62 moons. Neptune has only 13 known moons. The most recent moons found for each of these planets are irregular captured asteroids, often with radii of just a few kilometers. Their small size makes discovery difficult. This list of 27 moons (located on page 45), then, can only be thought of as a provisional list.

Neptune: Fast Facts about a Planet in Orbit

As soon as astronomers discovered Uranus in 1781 and began tracking it carefully, they saw that its orbit deviated from what was expected. This immediately brought them to the hypothesis that another even more distant planet was perturbing Uranus's orbit with its gravity. John Couch Adams was determined to find this more distant planet while still an undergraduate mathematics student at Cambridge University in England. He made careful calculations for two years based on Uranus's orbital perturbations, eventually predicting Neptune's position almost perfectly. The perfection of his calculations was not known at the time because he had great difficulty in getting any astronomer's attention. He tried for months to reach George Airy, Britain's Royal Astronomer. The story is frequently told that Airy eluded Adams and felt disdainful of him, but in fact Adams failed to make appointments before trying to see the busy Airy, and when Airy read Adams's calculations in the end and wrote asking for additional calculations, Adams never answered. Airy in turn did not use the telescope at the Greenwich Observatory to look for the new planet, and wrote to Adams that the observatory was too busy.

Meanwhile, Urbain Le Verrier, an astronomer at the Paris Observatory, was making the same calculations in France. He also failed to get the attention of the astronomical community, even after presenting his results at the French Academy of Sciences. When Airy wrote to Le Verrier asking the same questions he had of Adams, Le Verrier answered. Airy communicated with the Cambridge Observatory, where a search was begun but failed because of incomplete record making. Neptune was actually found but overlooked by the Cambridge group, much to their later distress.

In 1846 Le Verrier finally got the attention of Johann Galle and Heinrich d'Arrest, astronomers at the Berlin Observatory. They, along with observatory director Johann Encke, looked in the area of sky Le Verrier predicted. Over two nights, they saw a small greenish point of light move, and knew they had found the new planet. This discovery, motivated by the work of Le Verrier, was a serious embarrassment to Airy. This embarrassment was exacerbated by the long rivalry between the countries. Airy allegedly took an extended vacation to avoid the controversy.

Soon Le Verrier and Adams were announced as codiscoverers, and they continue to share that honor today. Naming the planet caused even more trouble: There was a motion in France to name the planet Le Verrier, begun by Le Verrier himself, while others suggested Couch, Janus, and Oceanus,

Symbol for Neptune

© Infobase Publishing

Many solar system objects have simple symbols; this is the symbol for Neptune.

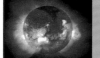

FUNDAMENTAL INFORMATION ABOUT NEPTUNE

As he did with Uranus, Galileo also saw Neptune in 1612, 234 years before its confirmed discovery. His telescope was not powerful enough to distinguish Neptune from the background stars. It is also reported that Galileo mistook Neptune for one of the Jovian moons. Neptune was discovered in 1846, and the planet takes 164.8 Earth years to orbit the Sun. Neptune has therefore made almost one entire orbit since the date of its discovery.

Like Jupiter and Saturn, Uranus and Neptune have enough similarities to be considered as a pair. The two are almost exactly the same size, as noted in the table below, though Neptune is far more dense than Uranus. The density of the two planets sets them apart from the strikingly low density of Jupiter and Saturn. Uranus and Neptune also share the distinctive patterns of their magnetic fields, along with the field strengths.

FUNDAMENTAL FACTS ABOUT NEPTUNE	
equatorial radius at the height where atmospheric pressure is one bar	15,388 miles (24,764 km), or 3.89 times Earth's radius (Uranus is 15,882 miles or 25,559 km)
polar radius	15,126 miles (24,442 km)
ellipticity	0.017, meaning the planet's equator is almost 2 percent longer than its polar radius
volume	1.50×10^{13} cubic miles (6.253×10^{13} km^3), or 57.7 times Earth's volume
mass	2.253×10^{26} pounds (1.024×10^{26} kg)
average density	110 pounds per cubic foot (1,760 kg/m^3), or 3.13 times less dense than Earth (Uranus is 79.4 lb/ft^3 or 1,270 kg/m^3)
acceleration of gravity on the surface at the equator	35.8 feet per second squared (11 m/sec^2), or 1.12 times Earth's gravity
magnetic field strength at the surface	2×10^{-5} tesla
rings	six
moons	13 presently known

but in the end the planet was named after Neptune, the son of Saturn. Each planet, and some other bodies in the solar system (the Sun and certain asteroids), has been given its own symbol as a shorthand in scientific writing. The planetary symbol for Neptune, the triton, is shown on page 66.

Though Adams and Le Verrier might be assumed to be archrivals, Adams in particular seems to have been a modest and companionable person. He even refused a knighthood that was offered him in 1847. Adams met Le Verrier in Oxford in June 1847. According to James Glaisher, president of the London Mathematical Society in the late 19th century and one of Adams's biographers, "He uttered no complaint, he laid no claim to priority, and Le Verrier had no warmer admirer." Le Verrier went on to detect discrepancies in the perihelion of Mercury's orbit that eventually helped prove Einstein's theory of relativity, though Le Verrier sought a (nonexistent) planet closer to the Sun than Mercury that he thought must cause Mercury's strange orbit.

Even after Neptune's proper discovery in the 19th century, scientists continued to look for another large planet because the mass calculated for Neptune could not explain all of the irregularities in its orbit, and to a lesser extent, in the orbits of Uranus and Saturn. Neptune's mass was finally recalculated using data from its *Voyager 2* encounter in 1989, and this new, more precise mass now accounts for all the discrepancies in the orbits of the planets, making it unlikely that another planet of any significant size will be found past Pluto.

Uranus and Neptune differ more in their orbital characteristics than in those of the physical planet. The two rotate at about the same rate (almost 18 hours for Uranus and about 16 hours for Neptune), though Neptune rotates with an inclination similar to the Earth's while Uranus appears to have been highly disturbed, probably by an impact, and now orbits with its rotation axis almost in the plane of its orbit. Neptune, of course, is much farther from the Sun. Uranus orbits at an average distance of 19 AU, while Neptune orbits at 30 AU. Both planets have near-circular orbits that lie very close to the ecliptic plane, entirely unlike their outer neighbor Pluto.

NEPTUNE'S ORBIT

rotation on its axis ("day")	16 Earth hours, seven minutes
rotation speed at equator	6,000 miles per hour (9,658 km/hour)
rotation direction	prograde (counterclockwise when viewed from above its north pole)
sidereal period ("year")	164.8 Earth years
orbital velocity (average)	3.404 miles per second (5.478 km/sec)
sunlight travel time (average)	four hours, nine minutes, and 59 seconds to reach Neptune
average distance from the Sun	2,795,084,800 miles (4,498,252,900 km), or 30.07 AU
perihelion	2,771,087,000 miles (4,459,630,000 km), or 29.811 AU from the Sun
aphelion	2,819,080,000 miles (4,536,870,000 km), or 30.327 AU from the Sun
orbital eccentricity	0.00859
orbital inclination to the ecliptic	1.77 degrees
obliquity (inclination of equator to orbit)	29.56 degrees

All orbits are ellipses, not circles. An ellipse can be thought of simply as a squashed circle, resembling an oval. Neptune's orbit is very close to circular, but it is still an ellipse. The proper definition of an ellipse is the set of all points that have the same sum of distances from two given fixed points, called foci. This definition can be demonstrated by taking two pins and pushing them into a piece of stiff cardboard, then looping a string around the pins (see figure on page 70). The two pins are the foci of the ellipse. Pull the string away from the pins with a pencil, and draw the ellipse, keeping the string taut around the pins and the pencil all the way around. Adding the distance along the two string segments from the pencil to each of the pins will give the same answer each time: The ellipse is

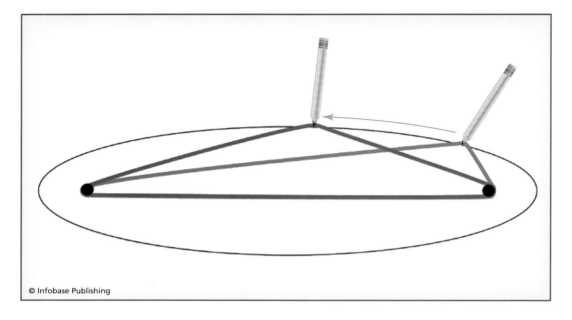

© Infobase Publishing

Making an ellipse with string and two pins: Adding the distance along the two string segments from the pencil to each of the pins will give the same sum at every point around the ellipse. This method creates an ellipse with the pins at its foci.

the set of all points that have the same sum of distances from the two foci.

The mathematical equation for an ellipse is

$$\frac{x^2}{a^2} + \frac{y^2}{b^2} = 1,$$

where x and y are the coordinates of all the points on the ellipse, and a and b are the semimajor and semiminor axes, respectively. The semimajor axis and semiminor axis would both be the radius if the shape was a circle, but two radii are needed for an ellipse. If a and b are equal, then the equation for the ellipse becomes the equation for a circle:

$$x^2 + y^2 = n,$$

where n is any constant.

When drawing an ellipse with string and pins, it is obvious where the foci are (they are the pins). In the abstract, the foci can be calculated according to the following equations:

Coordinates of the first focus

$$= (+\sqrt{a^2 - b^2}, 0).$$

Coordinates of the second focus

$$= (-\sqrt{a^2 - b^2}, 0).$$

An important characteristic of an ellipse, perhaps the most important for orbital physics, is its eccentricity: a measure of how different the semimajor and semiminor axes of the ellipse are. Eccentricity is dimensionless and ranges from 200 to one, where an eccentricity of zero means that the figure is a circle, and an eccentricity of one means that the ellipse has gone to its other extreme, a parabola (the reason an extreme ellipse becomes a parabola results from its definition as a conic section). One equation for eccentricity is

$$e = \sqrt{1 - \frac{b^2}{a^2}},$$

where a and b are the semimajor and semiminor axes, respectively (see figure below). Another equation for eccentricity is

$$e = \frac{c}{a},$$

where c is the distance between the center of the ellipse and one focus. The eccentricities of the orbits of the planets vary

Semimajor and Semiminor Axes, Foci

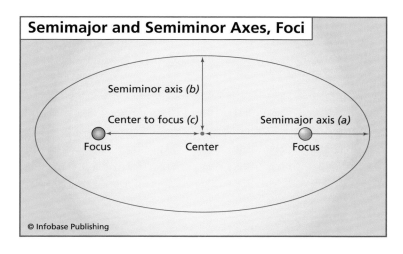

Semiminor axis (b)

Center to focus (c)

Semimajor axis (a)

Focus

Center

Focus

© Infobase Publishing

The semimajor and semiminor axes of an ellipse (or an orbit) are the elements used to calculate its eccentricity, and the body being orbited always lies at one of the foci.

Though the orbits of planets are measurably eccentric, they deviate from circularity by very little. This figure shows the eccentricity of the Earth's, Mercury's, and Pluto's orbits in comparison with a circle.

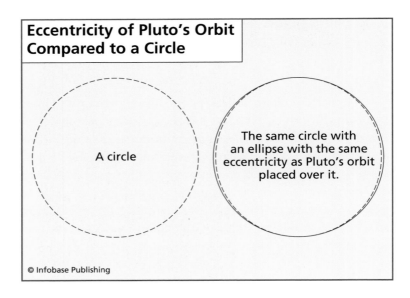

Eccentricity of Pluto's Orbit Compared to a Circle

A circle

The same circle with an ellipse with the same eccentricity as Pluto's orbit placed over it.

© Infobase Publishing

widely, though most are very close to circles, as shown in the figure above. Pluto has the most eccentric orbit, and Mercury's orbit is also very eccentric, but the rest have eccentricities below 0.09.

Johannes Kepler, the prominent 17th-century German mathematician and astronomer, first realized that the orbits of planets are ellipses after analyzing a series of precise observations of the location of Mars that had been taken by his colleague, the distinguished Danish astronomer Tycho Brahe. Kepler drew rays from the Sun's center to the orbit of Mars and noted the date and time that Mars arrived on each of these rays. He noted that Mars swept out equal areas between itself and the Sun in equal times, and that Mars moved much faster when it was near the Sun than when it was farther from the Sun. Together, these observations convinced Kepler that the orbit was shaped as an ellipse, not as a circle, as had been previously assumed. Kepler defined three laws of orbital motion (listed in the table on page 73), which he published in 1609 and 1619 in his books *New Astronomy* and *The Harmony of the World*. These three laws, listed in the table, are still used as the basis for understanding orbits.

KEPLER'S LAWS	
Kepler's first law:	A planet orbits the Sun following the path of an ellipse with the Sun at one focus.
Kepler's second law:	A line joining a planet to the Sun sweeps out equal areas in equal times (see figure on page 74).
Kepler's third law:	The closer a planet is to the Sun, the greater its speed. This is stated as: The square of the period of a planet T is proportional to the cube of its semimajor axis R, or $T \propto R^{\frac{3}{2}}$ as long as T is in years and R in AU.

While the characteristics of an ellipse drawn on a sheet of paper can be measured, orbits in space are more difficult to characterize. The ellipse itself has to be described, and then the ellipse's position in space, and then the motion of the body as it travels around the ellipse. Six parameters are needed to specify the motion of a body in its orbit and the position of the orbit. These are called the orbital elements (see the figure on page 76). The first three elements are used to determine where a body is in its orbit.

  a **semimajor axis** The semimajor axis is half the width of the widest part of the orbit ellipse. For solar system bodies, the value of semimajor axis is typically expressed in units of AU. Neptune's semimajor axis is 30.069 AU.

  e **eccentricity** Eccentricity measures the amount by which an ellipse differs from a circle, as described above. An orbit with $e = 0$ is circular, and an orbit with $e = 1$ stretches into infinity and becomes a parabola. In between, the orbits are ellipses. The orbits of all large planets are almost circles: The

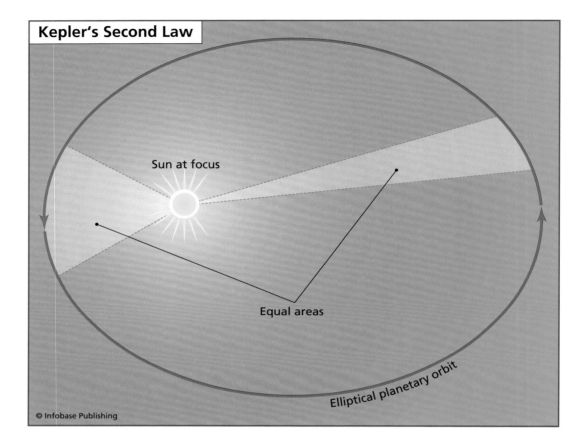

Kepler's Second Law

Sun at focus

Equal areas

Elliptical planetary orbit

© Infobase Publishing

Kepler's second law shows that the varying speed of a planet in its orbit requires that a line between the planet and the Sun sweep out equal areas in equal times.

Earth, for instance, has an eccentricity of 0.0068, and Neptune's eccentricity is 0.0113.

M **the mean anomaly** Mean anomaly is an angle that moves in time from zero to 360 degrees during one revolution, as if the planet were at the end of a hand of a clock and the Sun were at its center. This angle determines where in its orbit a planet is at a given time, and is defined to be zero degrees at *perigee* (when the planet is closest to the Sun), and 180 degrees at *apogee* (when the planet is farthest from the Sun). The equation for mean anomaly *M* is given as

$$M = M_0 + 360\left(\frac{t}{T}\right),$$

where M_o is the value of M at time zero, T is the orbital period, and t is the time in question.

The next three Keplerian elements determine where the orbit is in space.

i **inclination** In the case of a body orbiting the Sun, the inclination is the angle between the plane of the orbit of the body and the plane of the ecliptic (the plane in which the Earth's orbit lies). In the case of a body orbiting the Earth, the inclination is the angle between the plane of the body's orbit and the plane of the Earth's equator, such that an inclination of zero indicates that the body orbits directly over the equator, and an inclination of 90 indicates that

A series of parameters called orbital elements is used to describe exactly the orbit of a body.

Orbital Elements

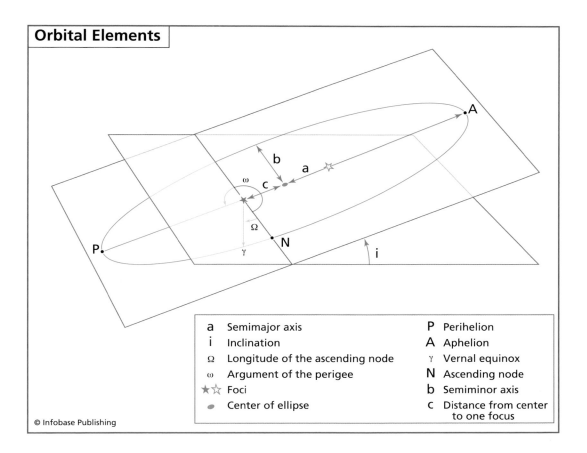

a	Semimajor axis	P	Perihelion
i	Inclination	A	Aphelion
Ω	Longitude of the ascending node	γ	Vernal equinox
ω	Argument of the perigee	N	Ascending node
★☆	Foci	b	Semiminor axis
⬤	Center of ellipse	c	Distance from center to one focus

the body orbits over the poles. If there is an orbital inclination greater than zero, then there is a line of intersection between the ecliptic plane and the orbital plane. This line is called the line of nodes. Neptune's orbital inclination is 1.769 degrees (see table below).

Ω **longitude of the ascending node** After inclination is specified, an infinite number of orbital planes is still possible: The line of nodes could cut through the Sun at any longitude around the Sun. Notice that the line of nodes emerges from the Sun in two places.

OBLIQUITY, ORBITAL INCLINATION, AND ROTATION DIRECTION FOR ALL THE PLANETS AND PLUTO

Planet	Obliquity (inclination of the planet's equator to its orbit; tilt); remarkable values are in italic	Orbital inclination to the ecliptic (angle between the planet's orbital plane and the Earth's orbital plane); remarkable values are in italic	Rotation direction
Mercury	0° (though some scientists believe the planet is flipped over, so this value may be 180°)	7.01°	prograde
Venus	*177.3°*	3.39°	*retrograde*
Earth	23.45°	0° (by definition)	prograde
Mars	25.2°	1.85°	prograde
Jupiter	3.12°	1.30°	prograde
Saturn	26.73°	2.48°	prograde
Uranus	*97.6°*	0.77°	*retrograde*
Neptune	28.32°	1.77°	prograde
Pluto	*119.6°*	*17.16°*	*retrograde*

One is called the ascending node (where the orbiting planet crosses the Sun's equator going from south to north). The other is called the descending node (where the orbiting planet crosses the Sun's equator going from north to south). Only one node needs to be specified, and by convention the ascending node is used. A second point in a planet's orbit is the vernal *equinox,* the spring day in which day and night have the same length (*equinox* means "equal night"), occurring where the plane of the planet's equator intersects its orbital plane. The angle between the vernal *equinox* γ and the ascending node N is called the longitude of the ascending node. Neptune's longitude of the ascending node is 131.722 degrees.

ω **argument of the perigee** The argument of the perigee is the angle (in the body's orbit plane) between the ascending node N and perihelion P, measured in the direction of the body's orbit. Neptune's argument of the perigee is 273.249 degrees.

The complexity of the six measurements shown above demonstrates the extreme attention to detail that is necessary when moving from simple theory ("every orbit is an ellipse") to measuring the movements of actual orbiting planets. Because of the gradual changes in orbits over time caused by gravitational interactions of many bodies and by changes within each planet, natural orbits are complex, evolving motions. To plan with such accuracy space missions such as the recent Mars Exploration Rovers, each of which landed perfectly in their targets, just kilometers long on the surface of another planet, the mission planners must be masters of orbital parameters. The complexity of calculating an actual orbit also explains the difficulty earlier astronomers had in calculating Neptune's, Uranus's, and Saturn's orbits. Their skill is apparent in the fact that only an incorrect value for Neptune's mass stood between them and reproducing the planet's orbits perfectly. The discrepancy stood for more than 100 years, until *Voyager 2* visited Neptune and other planets and took more precise measurements.

6

The Interior of Neptune

Though Neptune is 1.5 times farther from the Sun than Uranus is, its surface temperature is about the same as that of Uranus. Unlike quiescent Uranus, Neptune does release heat from its interior. Neptune's internal heat makes up for the lower solar flux it receives at its great distance from the Sun. Neptune receives about 2.6 times less solar heat than does Uranus. Neptune must produce 1.6 times as much heat internally as it receives from the Sun to maintain its surface temperature and measured heat flux from its interior. This ratio between internal heat production and heat received from the Sun is the highest for any planet.

Heat flux is the rate of heat loss per unit area. Knowing the heat flux out of the surface of a planet allows the calculation of internal temperatures for the planet. Because Uranus has no heat flux, its internal temperatures are unknown. Neptune's, however, can be calculated. The shallow interior layers of Neptune consist of hydrogen (H_2), helium (He), and methane (CH_4), similar to Uranus's. Most of Neptune's interior is thought to consist of a deep layer of liquid, a sort of internal ocean. This ocean above the core consists mainly of water with some methane and ammonia at a temperature thought to be about 8,500°F (4,700°C). Pressure keeps the material in a liquid form rather than allowing it to evaporate.

For reasons that are not yet understood, Neptune is about 50 percent more dense than Uranus. Neptune is thought to have a small rocky core, but its core is not likely to be larger than Uranus's, perhaps about the mass of the Earth. Based on heat measurements at the surface, Neptune's core is thought to reach temperatures of 9,300°F (5,100°C), but the source of this heat is unknown. On terrestrial planets internal heat is partly produced by decay of radioactive elements, but the most common heat-producing radioisotopes (potassium, uranium, and thorium, among others; see the sidebar "Elements and Isotopes" on page 20) are not common on gas giant planets. Jupiter produces heat by the condensation and internal sinking of helium, but Neptune is thought to be too depleted in helium for this process to proceed. The source of its unusual internal heat remains unexplained.

Planetary magnetic fields are all thought to be caused in the most basic sense by the movement of heat from inside the planet toward its surface. Heat can be transported by conduction, simply by moving as a wave through a material and into adjoining materials. When the bottom of a pot on an electric stove heats up, it is heating through conduction from the electric element. Heat can also move by *convection*. If a piece of material is heated, it almost always expands slightly and becomes less dense that its unheated counterpart. If the material is less dense than its surroundings and is able to flow, gravity will compel it to rise and the more dense surroundings to sink into its place. In a planetary interior, heat moving away from the core can cause the planetary interior to convect. Hot material from nearer the core will rise. When rising hot material reaches shallower levels, it can lose its heat to space through conduction or radiation, and when its heat is lost, the material becomes denser and then sinks again.

The combination of vigorous convection in a material that can conduct electricity with rotation within a planet creates a dynamo effect leading to a planetary magnetic field. If the fluid motion is fast, large, and conductive enough, then a magnetic field can be not only created by but also carried and deformed by the moving fluid (this process creates the Sun's highly complex magnetic field). The inner core rotates faster than the outer core, pulling field lines into itself and twisting them. Fluid welling up from the boundary with the inner core is twisted by the Coriolis effect, and in turn twists the magnetic field. In these ways, it is thought, the field lines are made extremely complex.

Though the magnetic fields of all the gas planets differ from those of the terrestrial planets in form and strength, Neptune's is perhaps the most different. Neptune's magnetic dipole is tilted away from the planet's axis of rotation by 47 degrees and offset from the center of the planet by about 0.55 times the planet's radius, as shown in the figure below. The field created is almost exactly as strong as Uranus's field at the level of one bar pressure, which is also the strength of Earth's field at its surface and about 10 times weaker than Jupiter's immense field.

On the terrestrial planets a dynamo is thought to form in the liquid metal outer core, convecting around the solid inner

Neptune's magnetic field is offset from both the center and rotation axis of the planet. Though it is not shown in this figure, all planetary magnetic fields are also pulled out into long magnetotails by the force of the solar wind.

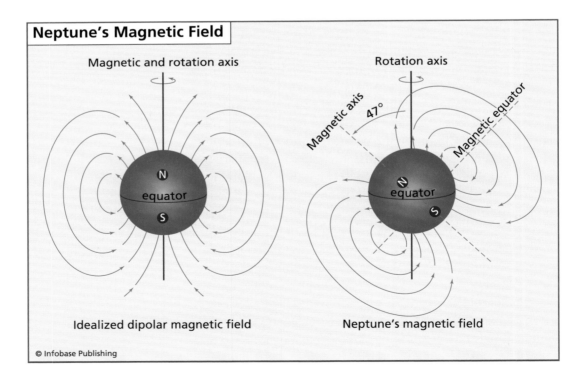

Neptune's Magnetic Field

Magnetic and rotation axis

Rotation axis

Magnetic axis

47°

Magnetic equator

N

equator

S

N

equator

S

Idealized dipolar magnetic field

Neptune's magnetic field

© Infobase Publishing

SABINE STANLEY AND PLANETARY MAGNETIC FIELDS

Sabine Stanley began by studying astronomy as an undergraduate at the University of Toronto, but she found that, despite the beauty of the field, she wanted to work on something a little closer and more tangible. Her freshman-year physics professor mentioned to her the work he had done during his postdoctoral study in Boston, and, after her junior year, Stanley began a summer job at Harvard University with Jeremy Bloxham.

Bloxham specializes in the study of planetary magnetic fields. This field can be a difficult one for a young student to begin with because it requires advanced understanding of the physics of magnetism and fluid mechanics, as well as a great facility with computer codes. Among other techniques, Bloxham creates models of dynamic planetary interiors to see how a magnetic field might be initiated, how it develops, and what it might look like today and in the future. Stanley's summer work went so well that after her graduation she returned to Harvard to do her doctoral work with Bloxham.

Stanley works in the general area of planetary magnetic fields, trying to answer questions about how different planets create their fields, how the shapes and strengths of the fields relate to the internal structure of the planets, and how magnetic field generation affects the planet itself. Does magnetic field generation speed the cooling of the planet's interior? Are there ways scientists can learn about a planet's past by studying its current magnetic field? She models the physical forces and interactions inside the fluid region of a planet using computer models. The models start with assumptions about the compositions and temperatures of the material inside the planet, and then they track the resulting movements of the material, its temperature changes, and the production of the magnetic field. These models have to solve a number of related equations (heat transport, conservation of mass, fluid motion in response to stress, and magnetic induction) for many individual locations over the planet for each of many time periods as the field develops. This scope of computing requires supercomputers; using a supercomputer (an SGI Origin 2000) with 16 processors, a single model must run continuously for several weeks to produce enough data to understand part of the development of a planetary field.

Stanley is trying specifically to understand what makes planets' magnetic fields different from one another. Though Uranus and Neptune are unusually similar planets,

(continues)

(continued)

their fields are quite different from each other in orientation and placement. The magnetic fields of the Earth, Jupiter, and Saturn are dipolar, similar to a bar magnet, but Uranus's and Neptune's fields are not. Their fields have strong quadrupole and octupole components and are not symmetric around their axes.

The Earth's magnetic field is generated in its fluid outer core, which is a thick shell of liquid metal with a small solid core consisting mainly of iron in its center. Uranus and Neptune, in contrast, may have small iron cores, but these cannot be the sources of their magnetic fields. Stanley and her colleagues think that the magnetic fields of Uranus and Neptune are generated in a shell of convecting fluid in the planets' interiors. Simple dynamo production in a thick fluid shell will create a dipolar magnetic field, according to theory and the models that Stanley runs. To create the complex magnetic fields of Uranus and Neptune, Stanley finds, a planet requires a thin shell of convecting fluid with nonconvecting fluid in its interior. The interior fluid may be nonconvecting because its density resists movements caused by heat transfer (this is called a stably stratified fluid). A thin shell of fluid can produce more complex magnetic field patterns because different areas of the shell have more trouble communicating with each other. Convective patterns are forced to be on a small scale because the shell is thin, and so more convection cells form over the area of the shell, and convection on one side of the shell is much more weakly related to convection on the other side than it would be in a deep fluid region with larger convection cells. A strong magnetic field, they find,

core. The gas giant planets lack any liquid metal layer; their magnetic fields are thought to form in circulating watery liquids. Jeremy Bloxham and Sabine Stanley at Harvard University have created models showing how the magnetic fields of Uranus and Neptune can be formed in thin convecting liquid shells in the planet's interiors, likely to consist, in a simple sense, of salty water (see the sidebar "Sabine Stanley and Planetary Magnetic Fields" on page 81). Their models demonstrate that the thin convecting shells in otherwise stable planetary interiors can create magnetic fields that are nonsymmetric and tipped away from the planet's axis of rotation.

People tend to think of magnetic fields just in terms of dipoles, meaning a two-poled system like a bar magnet. The Earth's magnetic field largely resembles a dipole, with mag-

can itself begin to move the stably stratified fluids beneath the convecting shell, creating even more complex magnetic field patterns.

On Uranus and Neptune, Stanley and her colleagues believe, the thin convecting shell that creates the magnetic dynamo consists of some combination of water, ammonia, and methane. Under high pressure the molecules of these materials break apart and form smaller molecules and atoms with electric charges. This ionic soup can act as electric currents as it moves because of convection. To test the predictions that their computer models make, though, Stanley and her colleagues require data from another mission to the distant gas giant planets. There is at least a discussion at NASA about a potential mission to Neptune that would measure its magnetic field in more detail.

Stanley finished her Ph.D. at Harvard in 2004 and moved on to a postdoctoral research position at the Massachusetts Institute of Technology (MIT), working with Maria Zuber. Zuber is a mission scientist for NASA as well as a professor of geophysics and the department head of the Earth, Atmospheric, and Planetary Science Department at MIT. With Zuber, Stanley focused on the magnetic field of Mercury to determine whether its anomalous field might be generated in a very thin shell of liquid metal around an otherwise solid core. In 2005 Stanley accepted faculty position at the University of Toronto. Stanley's work is at the forefront of the understanding of magnetic fields, and her work will continue to unravel some of the unanswered questions about planetary dynamics and evolution.

netic field lines flowing out of the south magnetic pole and into the north magnetic pole, but there are other, more complex configurations possible for magnetic fields. The next most complex after the dipole is the quadrupole, in which the field has four poles equally spaced around the sphere of the planet. After the quadrupole comes the octupole, which has eight poles. Earth's magnetic field is thought to degenerate into quadrupole and octupole fields as it reverses, and then to reform into the reversed dipole field. Neptune's field is largely a dipole but has strong high-order components, creating a complex field.

Electrical currents in shallow liquids and ices, rather than any generated by a core dynamo, may create Neptune's magnetic field. The strongest electric currents in Neptune are thought to exist at about half the radius of the planet. The tilt of the magnetic field relative to the rotation axis of the planet means that the shape of the *magnetosphere* (the region around Neptune devoid of solar wind because of magnetic field protection) changes dynamically as the planet rotates. This continuous complex change may contribute to heating of the highest portions of Neptune's atmosphere, described below. Both continuous and irregular, bursting radio emission come from Neptune, again very similar to Uranus's emissions. The rotation of the magnetic field is also thought to be a controlling producer of these emissions.

Thanks to the sophisticated computer modeling of the Harvard University team and others, some insight can be gained into the interior of this remote planet. With only density, heat flow, and magnetic field measurements, along with some compositional information from spectrometry, reasonable inferences can be gained without direct data. The understanding of Neptune is ahead of the understanding of Uranus simply because Neptune has measurable surface heat flow. All the scientists studying these distant planets, though, wish for a new space mission to take new measurements and allow testing of the hypotheses that have been made.

Surface Appearance and Conditions on Neptune

Neptune and Uranus are both blue planets because the trace of methane in their atmospheres absorbs red light, leaving only blue light to be reflected or emitted. Unlike its water-rich interior, Neptune's atmosphere consists of between 16 and 22 percent helium, about 2 percent methane, and the remainder hydrogen. A few other trace molecules are thought to contribute to cloud formation, including ammonia (NH_3), ammonium hydrosulfide (NH_4SH), and hydrogen sulfide (H_2S). Small amounts of water and carbon dioxide in Neptune's upper atmosphere seem to originate in interplanetary dust and meteorites. These delicate measurements have all been made with spectroscopic techniques from great distances, a testament to the ingenuity and persistence of the scientific community (for more, see the sidebar "Remote Sensing" on page 30).

Neptune has a *visual magnitude* of only 7.8, and so cannot be seen with the naked eye. The magnitude scale has no dimensions, but allows comparison of brightness between celestial objects. The lower the magnitude number, the brighter the object. The brightest star is Sirius, with a magnitude of –1.4. The full Moon has magnitude –12.7 and the Sun –26.7. The faintest stars visible under dark skies are around +6. During its close *opposition,* Mars rose to an *apparent magnitude* of −2.9. Neptune is therefore far dimmer than the faintest visible stars. Neptune's surface reflects almost a third of the sunlight that strikes it, so only about two-thirds of the paltry light that reaches distant Neptune is available to heat the planet. In the absence of internal heat, relying solely on heat from the Sun, Neptune's temperature at a pressure of one bar would be −375°F (−226°C). Because Neptune has an internal heat source, its mean temperature at a pressure of 1 bar is −328°F (−200°C). Beneath this level, temperatures rise with depth into the planet's interior, and above this level, temperatures also rise with height into the upper atmosphere.

Like Uranus, Neptune has an anomalously hot upper atmosphere. Uranus's atmospheric temperature reaches a high of perhaps 890°F (480°C) in the exosphere, the uppermost layer of its atmosphere. Neptune's exosphere appears to be slightly cooler, reaching only about 620°F (327°C). This heating is not well understood. Ultraviolet radiation from the Sun cannot heat the outer atmosphere to these temperatures. Though Neptune is much farther from the Sun, it has a slightly higher abundance of methane than does Uranus. Methane will help hold heat in the atmosphere and may contribute to additional solar heat retention on Neptune, but no atmospheric composition can explain the extremes of heat that have been measured.

The other gas giant planets have similarly hot exospheres, and on Jupiter the heat is thought to come from ionizing reactions tied to Jupiter's magnetic field and its auroras. Perhaps a similar process is at work on Uranus and Neptune. This process may not be able to explain the differences in temperature between Uranus and Neptune, though, because the planets have similarly strong magnetic fields and thus should accelerate ions to similar speeds and cause similar amounts of atmospheric heating. Other possibilities for exospheric heating are dissipation of gravity waves from the planet's interior, collision of supersonic jets of ions in the aurora, and other forms of heating caused by interactions between the atmosphere and the magnetic field.

Neptune has clouds, high winds, and large cyclonic storms that form and fade over months or years. The storms on Neptune are more distinct and common than those on Uranus. The variation of Neptune's surface can be seen in the image shown here, covering the entire surface of the planet. The location of the clouds is predicted based upon the temperature at which methane vapor will condense, and the levels and compositions are similar to those predicted and measured on Uranus. Methane ice clouds are expected to form at pressures less than about one bar. Between about five bars and one bar, ammonia and hydrogen sulfide ice clouds should form. At pressures greater than five bars, clouds of ammonium hydrosulfide, water, ammonia, and hydrogen sulfide form, both alone and in solution with one another. Water ice clouds are thought to form around 50 bars where the temperature is around 32°F (0°C). Neptune's deepest clouds, up to a few hundred bars pressure, are thought to consist of hydrogen sulfide (H_2S) and ammonia (NH_3).

Unlike Uranus, Neptune has a high-altitude, low-pressure layer of hydrocarbon haze above its methane ice clouds. Seran Gibbard of the Lawrence Livermore Laboratories and her colleagues have used high-resolution images and spectral analysis (breaking the light into its constituent wavelengths) from the 32-foot (10-m) Keck II Telescope to determine the pressures and compositions of atmospheric features on Neptune. Dr. Gibbard and her team find that most cloud features exist between 0.1 and 0.24 bars and that some bright features are caused by large downwellings of methane haze rather than by methane ice clouds.

Some winds and rotation patterns on Neptune have been measured at 1,200 miles per second (2,000 km/sec),

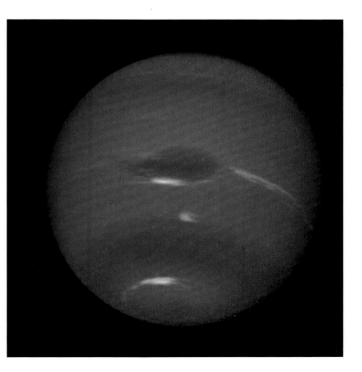

Neptune's image shows the Great Dark Spot and its companion bright smudge; on the west limb the fast-moving bright feature called Scooter and the little dark spot are visible. (NASA/JPL)

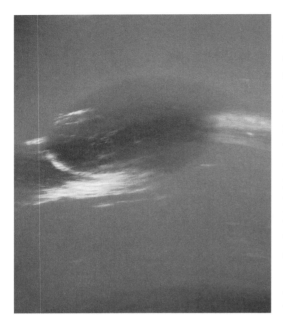

Neptune's Great Dark Spot in close-up (NASA/JPL)

by far the highest wind speeds in the solar system. Neptune's more usual wind speeds at the cloud tops are 1,300 feet per second (400 m/sec) at the equator and 800 feet per second (250 m/sec) nearer the poles, 5,000 or more times slower than the maximum measured winds. Wind speed contrasts of this scale do not exist on Earth. Neptune's circulation pattern is similar to though simpler than the zonal patterns (parallel to latitude) on Jupiter and Saturn. Neptune has large retrograde (westward) winds at the equator, and eastward winds poleward of about 50 degrees latitude.

Voyager 2 images revealed a dark oval storm on Neptune at about the same latitude as Jupiter's Great Red Spot around 30 degrees south latitude. This storm was named the Great Dark Spot and is shown in an image from *Voyager 2* that also contains small dark spots and a bright feature nicknamed Scooter. Over the course of 1989, the Great Dark Spot moved about 10 degrees closer to the equator. The Great Dark Spot, like the Great Red Spot, is a high-pressure, anti-cyclonic storm. Its size oscillated from 7,500 to 11,100 miles (12,000 to 18,000 km) in length and from 3,200 to 4,600 miles (5,200 to 7,400 km) in width over an eight-day period while *Voyager 2* was taking images. Smaller dark spots were also seen by *Voyager 2*.

The two images of Neptune (see figure on page 89) were taken by *Voyager 2* at a distance of about 7.5 million miles (12 million km) from Neptune. During the 17.6 hours between the left and right images, the Great Dark Spot, at 22 degrees south latitude (left of center), has completed a little less than one rotation of Neptune. The smaller dark spot, at 54 degrees south latitude, completed a little more than one rotation, as can be seen by comparing its relative positions in the two pictures. The velocities of the Great Dark Spot and the smaller spot differ by about 220 miles an hour (100 meters per second).

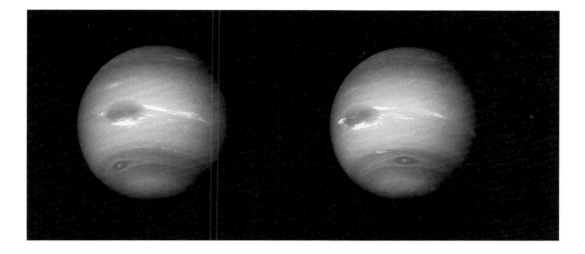

By 1994, images from the *Hubble Space Telescope* showed that the Great Dark Spot had disappeared. From 1994 through 1998, astronomers observed the development of new large dark spots in Neptune's northern hemisphere. The relatively rapid development and dissipation of large storms shows that Neptune's atmosphere is even less stable than Jupiter's. Though the composition of the dark material in the spots is unknown, bright attendant cirrus-like clouds are thought to be made of methane ice.

Neptune's measurable heat flow may well be tied to its extreme weather. In comparison with Uranus (very low surface heat flow), Neptune has bolder, faster, far more violent winds and weather. Since winds are driven by heating and cooling of the planet, the fastest winds in the solar system might be expected nearer the Sun, where solar heating can drive weather. That Neptune has the fastest measured winds in the solar system is therefore a surprise.

Neptune's surface was photographed twice 17.6 hours apart, showing the relative velocities of the spots in its atmosphere. (NASA/JPL/Voyager 2)

Neptune's Rings and Moons

Each new ring and moon system discovered in the solar system has brought its own unexpected results, requiring planetary scientists to revise their theories of formation and maintenance of planetary systems. Neptune is no exception. While several of Neptune's rings are conventional dusty features, some of its rings contain dense arcs connected by tenuously thin segments. The theory of ring formation requires an even distribution of particles around the planet, and so some complex and not completely understood interactions with small moons may be required to maintain these strange ring arcs. Neptune's moons contain an outstanding member, as well: its huge moon Triton. Triton is a moon of extremes: It is one of the brightest objects in the solar system, with an albedo of about 0.7; it is one of the very few moons with an atmosphere and one of only four solar system objects to have an atmosphere consisting largely of nitrogen; it orbits Neptune in a retrograde sense, unheard of for a major moon; and it has the coldest surface of any solar system object. Despite its exceptional cold, it has active cryovolcanic geysers (geysers of ice and perhaps exotic liquids), joining only Io and the Earth as bodies known to be volcanically active currently. Triton, with its seasonal heating, nitrogen atmosphere, and

Triton globe shows an icy coating in this image with a resolution of about 6 miles (10 km). In real color, the bright southern hemisphere of Triton is pink in tone, and the darker regions north of the equator pink or reddish. (NASA/JPL/ Voyager 2)

water content, is therefore a possible location for the development of life.

RINGS

Neptune's rings were first observed from Earth-based instruments in July 1984. William Hubbard, a scientist at the University of Arizona, and his colleagues André Brahic and Bruno Sicardy, L. Elicer, Françoise Roques, and Faith Vilas were observing the occultation of a bright star by Neptune. A star is said to be occulted by a planet if the planet passes directly in front of the star, such that the star's light passes through any atmosphere around the planet, or is briefly hidden by rings around the planet, before being completely obscured by the planet itself. Dips in the star's light intensity implied to the

In Neptune's outermost ring, material mysteriously clumps into three dense arcs separated by thinner ring material. (NASA/ JPL/Voyager 2)

scientists that there was an object nine miles (15 km) in width and 60 miles (100 km) long next to Neptune. The star's light was observed to dip only on one side of the planet and not the other, and so the objects around Neptune were inferred to be the solar system's first arcs, or incomplete rings.

These discoveries were so important that NASA and the Jet Propulsion Laboratory reprogrammed the *Voyager 2* mission specifically to look for the rings as it passed Neptune in 1989 (see figure above). Images from *Voyager 2*, such as those shown here, and more recently from the *Hubble Space Telescope*, show that most of the rings are complete and not arcs, but that they are narrow and dusty in areas and so will not reliably block starlight. They all seem to consist of ices with some silicates and carbonaceous materials.

Five rings around Neptune have been named. In order, moving away from Neptune, are the rings Galle, Le Verrier, Lassell, Arago, an unnamed ring, and Adams, as listed in the table above. Adams and Le Verrier are the narrowest and brightest of the rings, and those most likely to show in images of Neptune. Lassell is a broad ring, 2,485 miles (4,000 km) in width. Arago is the narrow bright edge of the ring Lassell, though not as bright as Adams and Le Verrier. Galle is another broad ring, with a width of about 1,200 miles (2,000 km). Of these six rings, only three or four may be distinct, independent entities. The unnamed ring is simply a trail of dust in the orbit of the moon Galatea, and so may not be a proper ring in terms of its formation and maintenance. Arago is simply the bright outermost edge of the broad ring Lassell.

The outermost ring, Adams, does consist of five distinct arcs: Liberté, Egalité 1, Egalité 2, Fraternité, and Courage, with a thin ring of dust connecting them. Liberté, the two Equalité arcs, and Fraternité are the brightest of the arcs, and make up between them about 40 degrees of arc. They are thought to consist of clumps of material bigger than those making up the rest of the Adams ring. The dim and dusty portions of the Adams ring appear to be about 19 miles (30 km)

NEPTUNE'S RINGS			
Ring	**Complete or arcs?**	**Radius (miles [km])**	**Width (miles [km])**
Galle	complete	26,000 (42,000)	1,200 (2,000)
Le Verrier	complete	33,000 (53,200)	<60 (<100)
Lassell	complete	34,300 (55,200)	2,500 (4,000)
Arago	complete	35,500 (57,200)	the bright outer edge of Lassell
unnamed	complete	38,270 (61,950)	a ring of dust in the moon Galatea's orbit
Adams	five arcs	39,100 (62,930)	nine (15)

(out of the plane of the ring), while the bright arc segments are 68 miles (110 km) in vertical extent.

The formation of rings almost requires that the particles become uniformly distributed along the entire ring. Particles at the inner edge of the ring orbit slightly faster than those in the outer portions of the ring because of their slightly closer proximity to Neptune. Ring particles are therefore moving past each other continually while orbiting. This action works to spread particles evenly around the ring.

Ring arcs, on the other hand, require some continuous outside influence to prevent their spreading out. Resonances (orbits in ratios of integers) may hold ring particles in specific portions of the ring. The moon Galatea has a 42:43 resonance with the Adams ring. The gravitational interactions between the moon and ring particles are magnified in specific positions when orbits are resonant, and so this resonance with Galatea may keep ring particles in arcs. More recent observations, however, indicate that the ring arcs are not in the correct position to be controlled by Galatea. In 1999 Christophe Dumas, from the Jet Propulsion Laboratory and the California Institute of Technology, and his colleagues looked at the rings using the *Hubble Space Telescope,* while another team led by Bruno Sicardy of the Observatory of Paris took images using the Canada-France-Hawaii telescope. Both teams concluded that Galatea's influence was insufficient to control the arcs. A second moon, perhaps as small as four miles (6 km) in diameter, could complete the effects required to maintain the arcs. Such a small moon would be very difficult to detect and so could be doing its work undiscovered.

MOONS

Neptune has 13 known moons, six of which were discovered by *Voyager 2,* and five of which were discovered in the last two years using high-resolution observations. Only Triton and Nereid were known before the space age. Many of Neptune's moons are irregular and therefore are unlikely to have differentiated internally or to have experienced any geologic activity. Even Nereid is irregular and unlikely to have

NEPTUNE'S MOONS

Moon	Orbital period [Earth days]	Moon's radius [miles (km)]	Year discovered	Orbital inclination [°]	Orbital eccentricity	Orbital direction
1. Naiad	0.294	30 by 19 by 16 (48 by 30 by 26)	1989	4.69	0.0003	prograde
2. Thalassa	0.311	34 by 31 by 16 (54 by 50 by 26)	1989	0.135	0.0002	prograde
3. Despina	0.335	56 by 46 by 40 (90 by 74 by 64)	1989	0.068	0.0002	prograde
4. Galatea	0.429	63 by 57 by 45 (102 by 92 by 72)	1989	0.034	0.0001	prograde
5. Larissa	0.555	67 by 63 by 52 (108 by 102 by 84)	1981/1989 (see text)	0.205	0.0014	prograde
6. Proteus	1.122	135 by 129 by 125 (218 by 208 by 201)	1989	0.075	0.0005	prograde
7. Triton	5.88	841 (1,353)	1846	156.8	0.000	retrograde
8. Nereid	360.1	106 (170)	1949	7.09	0.751	prograde
9. Halimede (S/2002 N1)	1,874.8	19 (30)	2002	112.7	0.264	prograde
10. Sao (S/2002 N2)	2,925.6	14 (22)	2002	53.5	0.137	prograde
11. Laomedeia (S/2002 N3)	2,980.4	14 (21)	2002	37.8	0.397	prograde
12. Psamathe (S/2003 N1)	9,136.1	12 (19)	2003	126.3	0.381	prograde
13. Neso (S/2002 N4)	9,007.1	19 (30)	2002	136.4	0.571	prograde

differentiated (see the sidebar "Accretion and Heating" on page 52). The exception is Triton, a large moon with geologic activity and an atmosphere.

The four moons closest to Neptune orbit within the Adams ring. Galatea orbits very close to the Adams rings, creating its own faint ring of dust. Larissa orbits just outside the ring system, and Proteus beyond Larissa. Even Proteus is so close to Neptune that Earth-based observatories cannot resolve it. The five moons discovered in 2002 and 2003 are small, and little is known about them.

1. Naiad

In Greek mythology, naiads are the water spirits who live in and govern springs, streams, and fountains. The moon Naiad is an irregular and relatively small satellite. Naiad's orbit is almost circular (eccentricity 0.0003) and lies close to Neptune's equator (orbital inclination 4.7 degrees). Naiad, Thalassa, Despina, Galatea, Larissa, and Proteus were all discovered by *Voyager 2,* with Naiad the last to be discovered.

2. Thalassa

Thalassa is the Greek word for "sea." In Greek mythology, Thalassa was the daughter of Ether, the god of the upper atmosphere, and his sister Hemera, the goddess of daylight, whose job it was to pull away the nighttime veils of the god Erebus. Thalassa, like Naiad, shows no sign of geological activity or alteration. Thalassa's orbit is even more perfectly circular than Naiad's (eccentricity 0.0002) and lies even closer to Neptune's equatorial plane (orbital inclination 0.135 degrees).

3. Despina

In Greek mythology, Despina is the daughter of Neptune and Demeter, the goddess of fertility. Despina is thought to be the inner shepherd moon to the Le Verrier ring. Like the moons interior to it, Despina shows no sign of geological activity or alteration. Its orbital eccentricity is 0.0002, and its orbital inclination is a mere 0.07 degrees.

4. Galatea

Galatea is named for a naiad in Greek mythology who lived on the island of Sicily and was the love interest of the cyclops Polyphemos, who imprisoned the hero Odysseus. Galatea is an inner shepherd to the Adams ring, with its curious arcs. Galatea was thought for a time to be responsible for the formation and maintenance of the ring arcs through gravitational interaction, but more thorough study has shown that one moon cannot be responsible alone. There may be a small, as-yet-unseen moon interacting with Galatea and the arcs. Galatea has a nearly circular orbit with very low inclination.

5. Larissa

In Greek mythology, Larissa was either the daughter or mother of Pelasgus, the founding citizen of the oldest group of inhabitants of Greece, the Pelasges, according to Herodotus. The existence of the moon Larissa was first implied by data from a stellar occultation experiment in 1981 led by Harold Reitsema, now a manager at Ball Aerospace. Larissa's existence was confirmed in images obtained in 1989 by *Voyager 2*. The slightly larger sizes of Larissa and Proteus (when compared to the inner moons) allows slightly better images to be taken. Like Proteus, Larissa is irregularly shaped and heavily cratered. Larissa orbits just outside the Adams ring, but it is not thought to influence the maintenance of Adams's arcs.

6. Proteus

Proteus is named after the mythological shape-shifting son of Poseidon, god of the sea, and Tethys, a titan and goddess of ocean fertility. (Proteus's family tree points out a fundamental inconsistency: Neptune is the god of the sea in Roman mythology, but all of Neptune's moons are named for figures in Greek mythology.) Proteus is Neptune's second-largest moon after Triton and is about as large as a solar system body can be without differentiating internally and obtaining a spherical shape. Its surface is exceptionally dark, reflecting only 6 percent of the sunlight that strikes it; Proteus is therefore one of the darkest solar system moons and difficult to see in images.

Its surface is heavily cratered. The largest crater is Pharos, with a diameter of 155 miles (250 km), forming a nine-mile (15-km)-deep depression that covers most of the moon's southern hemisphere.

7. Triton

Triton is the largest of Neptune's moons, at 1,682 miles (2,706 km) in diameter, and is named for a Greek god of the sea, son of Poseidon (Neptune in the Roman pantheon). William Lassell, a wealthy Liverpool brewer and owner of the largest telescope in Britain, discovered it in 1846, just a month after Neptune itself. Before the mission of *Voyager 2*, Triton was thought to be the largest moon in the solar system because it is so bright, with an albedo of 0.7. Triton's high albedo means that its surface reflects more than 10 times as much sunlight as does its dark neighbor Proteus. Its brightness was misleading,

This view of Neptune's moon Triton is about 300 miles (500 km) across and encompasses two depressions, possibly old impact basins, that have been extensively modified by flooding, melting, faulting, and collapse. (NASA/JPL/ Voyager 2)

though, because Ganymede, a satellite of Jupiter, is actually the largest moon in the solar system.

Triton orbits in the opposite direction from the planet's direction, and as a result of tidal friction is slowly spiraling in toward the planet. Triton is the only large moon in the solar system to have a retrograde orbit. Triton's orbit is almost circular but highly inclined, at 157 degrees to the plane of Neptune's equator. Because of its retrograde, highly inclined orbit, the moon is not stable and will eventually crash into Neptune; it is almost certainly a captured asteroid. Its proposed compositional similarity to Pluto also suggests that Triton was once in independent orbit around the Sun in the Kuiper belt, as Pluto is. Some gravitational interaction altered its orbit and allowed it to pass close enough for capture by Neptune. Its capture would have created gravitational disturbances with Neptune's other moons, and may be responsible for Nereid's highly eccentric orbit. Modeling indicates that Triton will collide with Neptune in 10 million to 100 million years, forming rings that may be larger than Saturn's.

Triton's atmosphere is its most remarkable feature; only two other moons in the solar system have atmospheres (Jupiter's Io and Saturn's Titan). Triton's atmospheric pressure at its surface is about 0.000014 bars. Its atmosphere consists mainly of molecular nitrogen (N_2) with about 1/100 of a percent of methane. Triton is one of only four solar system bodies that have a large nitrogen component in their atmospheres: The first is the Earth, the second is Saturn's moon Titan, and Pluto also has a sporadic, thin, nitrogen-based atmosphere. Though its atmosphere is exceptionally thin, it may still form thin clouds of nitrogen ices. Triton is the largest contributor of plasma to Neptune's magnetosphere, and nitrogen atoms from Triton are regularly added to Neptune's upper atmosphere.

Triton is also a moon of temperature extremes. Triton has the coldest known surface temperature of any planet or moon, about −390°F (−235°C). Only Triton's very low temperature allows it to retain its atmosphere. Triton's highly inclined orbit and Neptune's own obliquity cause Triton to experience complex and severe seasons, perhaps the most extreme of any solar system body. The moon also has bright polar caps, visible in this image of Triton's limb, consisting of condensed nitrogen and methane that wax and wane with the moon's seasons, much as do Mars's polar caps.

Triton is thought to consist of a mixture of ices and rocky material because its density is 131 pounds per cubic foot (2,066 kg/m³), midway between the densities of water at 63 pounds per cubic foot (1,000 kg/m³) and rock at about 190 pounds per cubic foot (~3,000 kg/m³). *Voyager 2* spotted active geysers up to five miles (8 km) high and trailing away 60 miles (100 km). Triton therefore is volcanically active, spewing what is thought to be a mixture of liquid nitrogen, dust, and methane or ammonia from ice volcanoes. There are few fresh impact craters on Triton because its surface has been re-covered with fresh volcanic effusions (in a small part of the equatorial region the moon is heavily cratered). Triton's old craters have all been extensively modified by volcanic flooding and flow of the crust, until they have become just suggestions of craters, called palimpsests,

Cantaloupe texture and faults on Triton are shown in this image taken by Voyager 2. The vertical linear feature in the image is probably a graben about 20 miles (35 km) wide. The ridge in the center of the graben may be ice that has welled up by plastic flow. The lack of craters in the surrounding terrain shows it is relatively young. (NASA/JPL/ Voyager 2)

Formation of a Graben

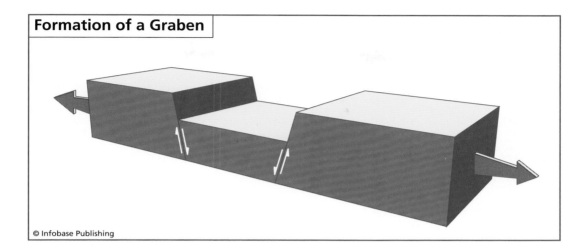

© Infobase Publishing

as shown in the figure on page 98. Triton, Io, and the Earth are the only solar system bodies known to be volcanically active at the present time, though Venus and Mars may also be volcanically active.

Triton's surface has a unique "cantaloupe" texture, crosscut with ridges between its poles and equator (see figure on page 100). This *Voyager 2* image shows the cantaloupe texture and two large crosscutting ridges, thought to be formed by faults. The smallest detail that can be seen is about 1.5 miles (2.5 km) across. Features on Triton's surface can be used to infer conditions inside the planet. Some of the troughs and ridges on Triton are thought to be *graben,* a geologic feature created by extension of the crust. Pulling the moon's crust in extension caused faults to form that allow blocks of the crust to sink down relative to their surroundings, as shown in the figure above. This kind of feature is relatively common on Earth and earned its name, *graben,* from an archaic German word for "grave." The long linear feature extending vertically across the image of cantaloupe texture probably is a graben about 20 miles (35 km) across. The ridge in the center of the graben is probably ice that has welled up by plastic flow in the floor of the graben. The surrounding terrain is a relatively young icy surface with few impact craters.

Two nine-mile (15-km)-wide troughs on Triton, called the Raz Fossae, are also thought to form a graben (for more on

Graben, long, low areas bounded by faults, are formed by crustal extension.

(continues on page 104)

FOSSA, SULCI, AND OTHER TERMS FOR PLANETARY LANDFORMS

On Earth the names for geological features often connote how they were formed and what they mean in terms of surface and planetary evolution. A caldera, for example, is a round depression formed by volcanic activity and generally encompassing volcanic vents. Though a round depression on another planet may remind a planetary geologist of a terrestrial caldera, it would be misleading to call that feature a caldera until its volcanic nature was proven. Images of other planets are not always clear and seldom include topography, so at times the details of the shape in question cannot be determined, making their definition even harder.

To avoid assigning causes to the shapes of landforms on other planets, scientists have resorted to creating a new series of names largely based on Latin, many of which are listed in the following table, that are used to describe planetary features. Some are used mainly on a single planet with unusual features, and others can be found throughout the solar system. Chaos terrain, for example, can be found on Mars, Mercury, and Jupiter's moon Europa. The Moon has a number of names for its exclusive use, including lacus, palus, rille, oceanus, and mare. New names for planetary objects must be submitted to and approved by the International Astronomical Union's (IAU) Working Group for Planetary System Nomenclature.

NOMENCLATURE FOR PLANETARY FEATURES

Feature	Description
astrum, astra	radial-patterned features on Venus
catena, catenae	chains of craters
chaos	distinctive area of broken terrain
chasma, chasmata	a deep, elongated, steep-sided valley or gorge
colles	small hills or knobs
corona, coronae	oval-shaped feature
crater, craters	a circular depression not necessarily created by impact
dorsum, dorsa	ridge
facula, faculae	bright spot
fluctus	flow terrain
fossa, fossae	narrow, shallow, linear depression

Feature	Description
labes	landslide
labyrinthus, labyrinthi	complex of intersecting valleys
lacus	small plain on the Moon; name means "lake"
lenticula, lenticulae	small dark spots on Europa (Latin for freckles); may be domes or pits
linea, lineae	a dark or bright elongate marking, may be curved or straight
macula, maculae	dark spot, may be irregular
mare, maria	large circular plain on the Moon; name means "sea"
mensa, mensae	a flat-topped hill with cliff-like edges
mons, montes	mountain
oceanus	a very large dark plain on the Moon; name means "ocean"
palus, paludes	small plain on the Moon; name means "swamp"
patera, paterae	an irregular crater
planitia, planitiae	low plain
planum, plana	plateau or high plain
reticulum, reticula	reticular (netlike) pattern on Venus
rille	narrow valley
rima, rimae	fissure on the Moon
rupes	scarp
sinus	small rounded plain; name means "bay"
sulcus, sulci	subparallel furrows and ridges
terra, terrae	extensive land mass
tessera, tesserae	tile-like, polygonal terrain
tholus, tholi	small dome-shaped mountain or hill
undae	dunes
vallis, valles	valley
vastitas, vastitates	extensive plain

(continues)

(continued)

The IAU has designated categories of names from which to choose for each planetary body, and in some cases, for each type of feature on a given planetary body. On Mercury, craters are named for famous deceased artists of various stripes, while rupes are named for scientific expeditions. On Venus, craters larger than 12.4 miles (20 km) are named for famous women, and those smaller than 12.4 miles (20 km) are given common female first names. Colles are named for sea goddesses, dorsa are named for sky goddesses, fossae are named for goddesses of war, and fluctus are named for miscellaneous goddesses.

The gas giant planets do not have features permanent enough to merit a nomenclature of features, but some of their solid moons do. Io's features are named after characters from Dante's *Inferno*. Europa's features are named after characters from Celtic myth. Guidelines can become even more explicit: Features on the moon Mimas are named after people and places from Malory's *Le Morte d'Arthur* legends, Baines translation. A number of asteroids also have naming guidelines. Features on 253 Mathilde, for example, are named after the coalfields and basins of Earth.

(continued from page 101)

words like *fossae*, see the sidebar "Fossa, Sulci, and Other Terms for Planetary Landforms" on page 102). Javier Ruiz, a scientist at the University of Madrid, has made an analysis of Triton's *lithosphere* using the dynamics implied by the fossae. If the stiff outer shell of the moon, its lithosphere, consists mainly of water and ammonia ices, then the depth and width of the fossae imply that the moon's lithosphere is only about 12 miles (20 km) thick. Beneath this ice layer may lie a shallow liquid ocean, perhaps similar to the ocean on Europa that is thought to be a potential site for the development of life.

8. Nereid

Nereid is named for the nereids, a group of about 50 sea nymphs in Greek mythology, the daughters of Nereus, a shape-changing sea god and oracle, and Doris, a titan. Nereid is, on average, about 15 times farther from Neptune than is Triton. Nereid takes 360 Earth days to make an orbit of Neptune, similar to the length of Earth's year. By comparison, Triton takes less than six Earth days to orbit Neptune. Nereid's highly eccentric orbit takes the moon from 834,210 miles (1,342,530 km) to 6,006,870 miles (9,667,120 km) away from Neptune. This is the most eccentric orbit of any moon in the solar system, with an eccentricity of 0.751. Nereid may be a captured asteroid, with its irregular shape and a highly eccentric orbit, or its orbit may have been disturbed when Neptune captured Triton.

The outermost five of Neptune's moons are newly discovered, small, irregular bodies about which little is known. In coming years they will be better imaged, no doubt, and studied, and still more remote and smaller moons will be discovered. The outermost moons now known orbit at three to 10 times the distance of Nereid to Neptune. All orbit in a prograde sense. They are likely to be captured asteroids, as are the outer moons of Jupiter and Saturn.

PART TWO
PLUTO AND THE KUIPER BELT

The Discovery of Pluto and the Kuiper Belt

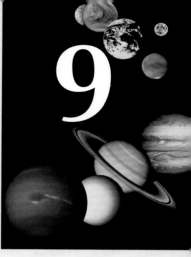

After the discovery of Neptune in 1846, astronomers soon predicted and began searching for a ninth planet based on perceived irregularities in the orbits of Neptune, Uranus, and even Saturn. The astronomers William Henry Pickering and Percival Lowell, sometimes partners and sometimes opponents, were among the most determined. They each used orbital analysis to predict the existence of a ninth planet, which Pickering called "Planet O" and Lowell called "Planet X." Pickering's accurate prediction was made in 1919, and the search began. (This Pickering should not be confused with a second William H. Pickering, who later became a towering figure in planetary science, launching the first U.S. satellite and directing the Jet Propulsion Laboratory for years.)

Lowell's first known prediction of a ninth planet was made in 1902. He began his search and orbital calculations soon after. Lowell conducted intense searches for the planet, but it was still undiscovered when he died in 1916. He was deeply discouraged by his failure to find the planet he was convinced existed. Lowell Observatory honored the memory of its founder in 1929 by hiring an amateur astronomer, Clyde Tombaugh, to continue the search for Planet X. Tombaugh was only 24 at the time and had never been to college. He was self-educated in astronomy and had built his own telescope. When Tombaugh wrote to the Lowell Observatory for encouragement in his passion, he was hired.

Tombaugh was exceptionally systematic in his search, far more so than Lowell had been. The technique he used was to take photos of the same section of sky on two successive nights and blink them on and off in a special viewing machine called a comparator. Stars appear stationary when the two photos are interchanged, but planets appear to move. Scanning photos in a comparator took great concentration; some photos of the Milky Way have up to a million stars in them.

On February 18, 1930, Tombaugh found the planet. He recognized it immediately and did quick calculations that showed its location was a billion miles past Neptune. The find was announced on March 13, 1930, the 149th anniversary of Herschel's discovery of Uranus and what would have been Lowell's 78th birthday. Because Pluto requires 247.9 Earth years to complete one orbit around the Sun, it has only completed about one-third of one orbit since its discovery.

The planet was named Pluto after the suggestion of an 11-year-old named Venetia Burney, who likened the planet's existence in perpetual darkness and cold to the life of the Roman god of the underworld. She reportedly suggested the idea at the breakfast table and her grandfather telegraphed the idea to the observatory, where they voted on names and chose hers. When Lowell calculated the position of Pluto based on the perturbations in Uranus's orbit, he estimated that the mass of the then-undiscovered Pluto was about 10 times the mass of the Earth. As it turns out, the Pluto-Charon system has a mass of only 0.0024 times the mass of the Earth, an error of three orders of magnitude. Lowell's calculations therefore cannot have been correct in even a rough sense, and so it is now thought that Tombaugh's discovery of the planet was just a happy accident.

Tombaugh died on January 17, 1997, just weeks before his 91st birthday (February 4). He had had a long and successful career, including major searches between 1977 and 1985 for other planets using the Palomar 48-inch (121.9-cm) reflecting telescope. These later searches were inspired in part by the idea that Uranus and Neptune had odd orbits due to the gravitational attraction of an as-yet-unfound Planet 10. Later it was discovered that Uranus and Neptune had completely regular orbits, but that the few data points available because of their slow movement as observed from Earth had made the orbital calculations undependable. When *Voyager 2* passed though the outer solar system in 1989 its data showed that a miscalculation in Neptune's mass is mainly responsible for the perturbations in Uranus's orbit.

The discovery of Pluto is tied inexorably to the theory and discovery of the Kuiper belt, a region of many small bodies orbiting the Sun with perihelia between 30 and 49 AU. Pluto's perihelion is at 29.658 AU, and, as is now understood, Pluto is simply one of the largest of this community of bodies. Pluto is now properly known as the first discovered Kuiper belt object, and so it is discussed here within the section on the Kuiper belt.

In 1950 Jan Oort, a Dutch astronomer from the University of Leiden, calculated that long-period comets (those with orbital periods of 200 years or more) had to come from a spherically distributed population of comets lying at 10,000 AU or more from the Sun. For some time, scientists assumed that short-period comets were bodies that had been perturbed out of the theoretical Oort cloud into orbits closer to the Sun. In 1972 Paul Joss, a professor of theoretical physics at the Massachusetts Institute of Technology, pointed out that a mass the size of Jupiter would be required to bring comets in from the Oort cloud and stabilize them in smaller orbits, and the probability of capture by Jupiter is prohibitively small.

A further problem emerged: Long-period comets have orbits that are roughly randomly oriented in space. Few orbit within or near the ecliptic plane. The orientations of their orbits led to the assumption that the Oort cloud is a spherical region around the otherwise approximately planar solar system. Short-period comets, however, orbit with 30 degrees of the ecliptic plane. They should originate, therefore, from a population with a similarly limited range of orbital inclinations.

In 1943 an Irish engineer and economist named Kenneth Edgeworth had postulated that short-period comets originated in a population of small bodies past Neptune. He wrote a paper in the *Journal of the British Astronomical Association* titled "The Evolution of Our Planetary System" in which he argued that there was no reason for the solar system to suddenly stop at the end of the planets. Smaller bodies should have accreted from the scanty material further out, he reasoned, and they would occasionally visit the inner solar system as comets. This hypothesis was published seven years before Oort postulated the existence of the larger and far more distant Oort cloud, and eight years before the famous Dutch astronomer Gerald Kuiper himself postulated the existence of the belt of small bodies.

Edgeworth was a minor figure in astronomy. He held no university position, though he published several papers on star formation. He spent much of his professional life as a military engineer abroad. Kuiper, on the other hand, was an established and respected astronomer in the university research community. In 1944 he reported the first proof of an atmosphere around Saturn's moon Titan; in 1948 he discovered Uranus's moon Miranda; and in 1949 he discovered Neptune's moon Nereid.

When in 1951 Kuiper hypothesized about the location of the short-period comets, his paper was widely read. Kuiper estimated how much mass had been lost from the inner solar system based on how much the planets' compositions differed from the solar composition. He assumed this mass must still exist in the solar system, and postulated that it existed outside Neptune, since it is physically unlikely that the solar nebula had a sharp outer edge. Edgeworth's paper, on the other hand, had been effectively lost from the literature (that is, not widely referenced and largely unknown and unread) in part because Kuiper himself neglected to refer to it. The reason for Kuiper's omission is not known, but subsequent scientists have acknowledged Edgeworth's contribution. Now his earlier hypothesis is credited as it deserves, and this population is often referred to as the Edgeworth-Kuiper belt or the Edgeworth-Kuiper disk.

Finally, in 1988, computer power increased to the point that simulations of cometary capture from the Oort cloud could be performed. Three scientists from the University of Toronto, Martin Duncan (now at Queen's University in Ontario), Tom Quinn (now at the University of Washington), and Scott Tremaine (now at Princeton University) proved Joss's theory about the rarity of gravitational capture from the Oort cloud. They further demonstrated that comets captured from the Oort cloud would keep an approximation of their original orbital inclinations. Comets captured from the Oort cloud, in other words, would have orbits at all inclinations to the ecliptic. Short-period comets therefore could not be derived from the Oort cloud but had to reside in their own separate population. These three scientists named their postulated population of bodies "Beyond Neptune the Kuiper belt," based on Gerald Kuiper's hypothesis.

Just before the conclusive computer modeling of Duncan, Quinn, and Tremaine, David Jewitt and Jane Luu of the University of California at Berkeley (now at UCLA and MIT, respectively) began an ambitious multiyear search for objects in the outer solar system. Using the large telescopes of the Mauna Kea observatory, the scientists examined minute sections of the night sky one after the other for small moving bodies at the edge of the known solar system. After five years of fruitless searching, they found the first Kuiper body in August 1992. Its preliminary minor planet designation was 1992 QB$_1$, and it is about 150 miles (240 km) in diameter. It was the first object found beyond Neptune since Pluto was found in 1930. Suddenly the decades of theory were finished, and humankind had proof that the solar system did not end with Pluto and Charon alone.

Pluto, Charon, and the tens of thousands of other members of the Kuiper belt appear to be remnants of the earlier formation processes of the solar system. They appear to be largely unaltered pieces of the original solar nebula, with the exception of some surface radiation damage and some radiogenic internal heating.

Kuiper belt bodies are often in resonance with Neptune, meaning that the Kuiper belt body's orbit and Neptune's orbit are related by an integer ratio (for example, the Kuiper belt body orbits three times for every two times Neptune completes an orbit). The accumulation of so many Kuiper belt bodies in resonant orbits may imply that Neptune and other gas giant planets migrated outward early in solar system formation, capturing bodies in resonant orbits as they moved. The Kuiper belt is also less massive than models indicate it should be, so it may represent the outcome of planetary migration resulting in both capture in resonant orbits and expulsion of mass from the solar system. The early Kuiper belt may also be analogous to the dust rings seen around young stars elsewhere in the galaxy. The Kuiper belt is thus a window into both early solar system formation and the formation of other planetary systems.

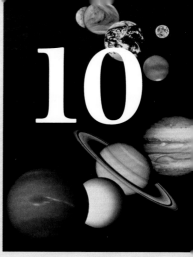

Pluto: Fast Facts about a Dwarf Planet in Orbit

Pluto is smaller than seven of the solar system's moons: Earth's Moon, Io, Europa, Ganymede, Callisto, Titan, and Triton. It is two-thirds the size of Earth's Moon, but 12,000 times farther away. The tiny sizes of and great distances to Pluto and Charon make taking good images almost impossible. One of the best *Hubble Space Telescope* images available of the two is shown on page 116. Charon is so difficult to distinguish that it was not discovered until 1978.

Pluto's mass is not accurately known because of the difficulty of discriminating between the masses of Pluto and Charon: The two orbit around their common center of mass and act more as binary planets than as a planet and its moon. The total mass of the Pluto-Charon system is estimated at 0.0024 Earth masses, but its division between Pluto and Charon is not well constrained, and therefore the densities of the two bodies are equally poorly constrained. Cathy Olkin and her colleagues at the Lowell Observatory have made the best and most recent estimate. Based on fine measurements of the motion of Pluto and Charon taken using the *Hubble Space Telescope* over a number of days in 2003, the team estimates that Charon contains about 11 percent of the total mass of the system and Pluto contains the remaining 89 percent, with

In this highest-resolution Hubble Space Telescope image, Pluto is the larger body to the lower left, and Charon is at upper right. (NASA/HST, NSSDC STScI-PR94-17)

errors of about 1 percent. All the data given on the two bodies here, including masses, volumes, and radii, must be considered estimates, with possible errors ranging from 1 percent to as much as 10 percent.

Each planet, and some other bodies in the solar system (the Sun and certain minor planets), has been given its own symbol as a shorthand in scientific writing. The symbol for Pluto is shown on page 119.

Pluto's orbit is unusually eccentric, moving from 29.7 AU from the Sun at perihelion to 49.85 AU at aphelion. Pluto was most recently at perihelion during 1989. During about two decades of each orbit, Pluto's eccentric path carries it inside the orbit of Neptune, which has a perihelion distance of 29.8 AU. Pluto was most recently closer to the Sun than Neptune between 1979 and 1999, a period that naturally encompasses Pluto's perihelion. Pluto's and Neptune's orbits are constrained in a 3:2 resonance, meaning that for every three times Neptune orbits the Sun, Pluto orbits twice. This resonance creates stable and long-lasting orbits.

Pluto's orbit is also highly inclined to the ecliptic plane (by 17.1 degrees). Its inclination carries Pluto from eight AU above the ecliptic plane to 13 AU below it. When Pluto is closest to

FUNDAMENTAL INFORMATION ABOUT PLUTO

The following table makes clear how small and poorly known Pluto is. Though it is considered a dwarf planet, far more is known about various moons than is known about Pluto. Pluto is smaller than all of Jupiter's Galilean satellites (Io, Europa, Ganymede, and Callisto) as well as Saturn's Titan, Uranus's Titania and Oberon, and Neptune's Triton. Pluto is hardly larger than four or five other Kuiper belt and possible Oort cloud objects already found.

FUNDAMENTAL FACTS ABOUT PLUTO	
equatorial radius	715 miles (1,151 km), or 0.18 times Earth's radius (though estimates for Pluto's radius vary by about 3 percent)
ellipticity	~0, but cannot be determined with present data more accurately than within 10 percent error
volume	1.53×10^9 cubic miles (6.39×10^{10} km^3), or 0.0059 times Earth's volume
mass	2.84×10^{22} pounds (1.29×10^{22} kg), or 0.0022 times Earth's mass
average density	111 pounds per cubic foot (1,750 kg/m^3) according to a NASA report, but other estimates range from 102 to 131 lb/ft^3 (1,600 to 2,060 kg/m^3)
acceleration of gravity on the surface at the equator	1.9 feet per second squared (0.58 m/sec^2), or 0.06 times Earth's gravity
magnetic field strength at the surface	unknown, but unlikely to have a magnetic field
rings	zero
moons	three

the Sun and inside Neptune's orbit, it is also about eight AU above the ecliptic. Between Pluto's resonance with Neptune and its movement out of the ecliptic plane it is never closer than 17 AU to Neptune; their closest approach happens when Pluto is near aphelion. In fact, Pluto makes closer (12 AU) and more frequent approaches to Uranus. Characteristics of Pluto's orbit are listed in the table below.

Pluto's orbital axis lies almost in its orbital plane, similar to but even more extreme than Uranus's obliquity. Pluto has the highest obliquity of any solar system body and as a result experiences large variations in solar heating over the course of its highly eccentric orbit. Pluto's high eccentricity, inclination,

PLUTO'S ORBIT	
rotation on its axis ("day")	6.39 Earth days
rotation direction	retrograde (clockwise when viewed from above the North Pole); the opposite direction to Earth's rotation
sidereal period ("year")	247.9 Earth years
orbital velocity (average)	2.95 miles per second (4.74 km/sec)
sunlight travel time (average)	five hours, 27 minutes, and 59 seconds to reach Pluto
average distance from the Sun	3,670,050,000 miles (5,906,380,000 km), or 39.482 AU
perihelion	2,756,902,000 miles (4,436,820,000 km), or 29.658 AU from the Sun
aphelion	4,583,190,000 miles (7,375,930,000 km), or 49.305 AU from the Sun
orbital eccentricity	0.2488, the highest in the solar system
orbital inclination to the ecliptic	17.14 degrees
obliquity (inclination of equator to orbit)	122.5 degrees, the highest of any planet in the solar system, but similar to Uranus's obliquity

Symbol for Pluto

© Infobase Publishing

Many solar system objects have simple symbols; this is the symbol for Pluto.

and obliquity all mark it as significantly different from the other, larger planets. From these characteristics alone Pluto might have been suspected to be a different type of solar system body, even before the Kuiper belt was discovered.

All the modern observations of Pluto have occurred when Pluto was at or very near its perihelion. Understanding of the planet is therefore based on measurements taken when the planet is at its warmest, and, in particular, atmospheric and surface characteristics may be different when the planet is at aphelion.

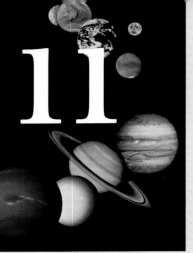

11

What Little Is Known about Pluto's Interior and Surface

Pluto's density is estimated at about 130 pounds per cubic foot (2,000 kg/m^3), though because density is the ratio of mass to volume, the error in Pluto's density estimate necessarily contains the errors of the estimates of the mass of the planet and its size compounded. A density in this range implies that the planet is between half rock and half ices and about 70 percent rock and 30 percent ices, similar to Triton's rock-to-ice ratio. The ratio of rock to ice depends, of course, on the compositions and therefore the densities of the rocks and ice themselves.

Water ice is thought to be the most abundant ice making up the planet because oxygen is more abundant in the solar system than nitrogen or carbon, the other ice-making atoms, and Pluto is expected to follow the solar system abundances. Water ice metamorphoses into different crystal structures depending on pressure: Humankind is most familiar with ice I, but with greater pressure ice transforms into ice III, ice V, and ice VI (see figure on page 121). Simultaneous changes in temperature allow the formation of other ice phases. These are called polymorphs: They consist of the same materials but have different crystal structures. At the pressure and temperature ranges of Pluto and Charon, water ice in their interiors may be in the forms of ices I, II, III, V, and IV.

At low pressures the water molecules organize themselves according to charge. The two hydrogen atoms in a water molecule are slightly positively charged, and the oxygen is

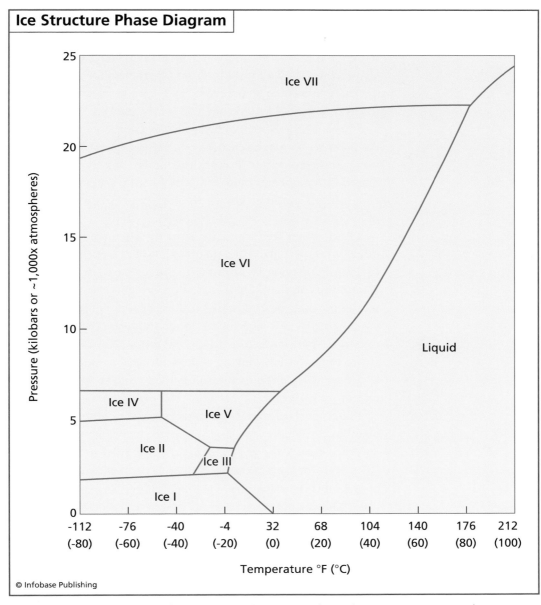

Ice Structure Phase Diagram

Pressure (kilobars or ~1,000x atmospheres)

Temperature °F (°C)

© Infobase Publishing

Water freezes into a variety of different crystal structures, depending on its pressure and temperature conditions.

slightly negatively charged. A hydrogen from a neighboring water molecule will therefore weakly bond with the oxygen, and the hydrogens will themselves weakly bond with oxygens in other molecules. Because water molecules are shaped like boomerangs, with the oxygen at the bend in the boomerang and a hydrogen at each end of the boomerang making an angle of about 108 degrees, the water molecules make a honeycomb shape when they all weakly bond together to form ice I.

The honeycomb structure of ice I is weak, not only because the water molecules are not efficiently packed together, but also because the electrical bonds between molecules are weak. As pressure is increased on ice I beyond the strength of the weak intermolecular bonds, the molecules eventually are forced into more and more efficient packing schemes. Pressure also inhibits melting, and so the higher-pressure polymorphs of ice can exist at temperatures above 32°F (0°C).

Pluto's other ices are likely to be nitrogen ice, carbon monoxide, and methane. The composition of the rock component in Pluto's interior is unknown. There may also be organic compounds on the planets, based on evidence from comet Halley. Pluto may be internally differentiated, in which case it would consist of a layer of ice about 190 miles (300 km) thick, underlain by a rock core about 560 miles (900 km) in radius. If the temperatures during Pluto's formation did not allow it to differentiate, then its interior is likely to be a heterogeneous mix of rock and ice.

Pluto's surface shows the most contrast of any solar system object aside from Iapetus and the Earth. Its brightness varies by 30 percent over the period of its rotation (just over six days). Alan Stern of the Southwest Research Institute and Marc Buie of Lowell Observatory have mapped Pluto's surface for the first time. The results were released March 7, 1996. The scientists took images in blue light over a period of several days using the *Hubble Space Telescope*, from which they made blurry maps of Pluto's surface as shown in the figure on page 128. During this period Pluto was 3 million miles (4.8 million km) from Earth.

The two upper images in the figure on page 128 show the actual *Hubble Space Telescope* data, and the lower images show

(continues on page 128)

HOW THE DISCOVERY OF ERIS CAUSED PLUTO TO LOSE ITS STATUS AS A PLANET

The discoveries of Kuiper belt objects in the mid-1990s led to increasing discussion in the scientific community about the definition of a planet. Pluto is smaller than several planetary moons and is surrounded by the Kuiper belt, now known to contain at least one body larger than Pluto. How is a planet defined? One working suggestion in the community of astronomers at the time was that a planet is anything that orbits the Sun and has a radius of at least 620 miles (1,000 km). This rule produces the usual nine planets. The 620-mile (1,000-km) radius law is arbitrary, though; is this an appropriate law to use as more and more planets are found outside this solar system? If a planet is required to have a moon, then Mercury and Venus are eliminated. If a planet is required to have an atmosphere, then Mercury is out again. If the object is simply required to orbit a star, then all the asteroids and comets become planets. In fact, asteroids are now commonly known as minor planets.

In a system based on size, Jupiter and Saturn might be defined as failed stars, Uranus and Neptune as planets, and everything smaller as minor planets. In an article in *Sky and Telescope* magazine, Alan Stern and Hal Levison presented this last system as an example of the extremes of this pursuit of a system of names. They went on to suggest a sensible system with two main criteria: The body cannot be massive enough that at any time it has generated energy internally by a fusion reaction (because then it would be a star of one sort or another); however, it must be massive enough that its shape is determined by gravity (in other words, it becomes spherical or ellipsoid through self-gravitation). This pair of criteria allows bodies with masses from one one-thousandth the mass of the Earth to 10,000 times the mass of the Earth to be defined as planets, and it also allows about 20 solar-system objects to be defined as planets, including the Moon and several others, the asteroids Ceres, Vesta, Pallas, and the Kuiper belt objects Quaoar, Ixion, and Varuna, among other bodies.

Michael Brown, a professor at the California Institute of Technology, and his colleagues then suggested another attribute required to earn the title planet: The body must orbit the Sun alone rather than in a population of like objects with like orbits. Ceres, Vesta, and Pallas cannot be planets because they orbit in the population of asteroids in the main belt. Quaoar, Ixion, and Varuna cannot be planets because they orbit in the Kuiper belt, and thus, Pluto is no longer considered a planet either. The list of planets in the solar system would be reduced by one in this scheme, and all bodies that

(continues)

(continued)

The IAU announcement that Pluto was losing its status as a planet received attention from the public.

are round through self-gravity but do not orbit alone would be known as planetoids. As Brown pointed out, Ceres and the other initially discovered asteroids were at first classified as planets until it was realized that they were a part of a larger population, and then they were demoted from planet status. By this reasoning, it is time, they argued, to demote Pluto because of the discoveries of so many new Kuiper belt bodies.

Despite the debate in the press and on the Internet, the International Astronomical Union issued a press release in 1999 stating unequivocally that it was not considering removing Pluto from the list of major planets (it is the smallest and most asteroidlike of the planets) and that it considered the matter of redefining planets closed.

On January 5, 2005, Mike Brown, Chad Trujillo, and David Rabonowitz discovered a very slow-moving, very distant object in the Kuiper belt. This object, temporarily labeled 2003 UB313 because its first observation, which had been missed by computer analysis because it was moving so slowly, was taken in 2003. The object immediately took on an informal name, Xena (convenient for starting with the letter *x*, as in Planet X), and its tiny moon became known as Gabrielle. The immediate question was: How big is Xena?

Because the size of an object in the Kuiper belt is judged in part by its brightness, the albedo of the object (how much light is reflected from its surface) must be known before brightness can be used to judge size. For any reasonable albedo they chose, however, this new object was going to be bigger than Pluto. This was the largest object to be discovered in the solar system in more than 150 years, and it was larger than one of the planets.

As better measurements of the object were made, the researchers determined that it is about 1,500 miles (2,400 km) in diameter, about 5 percent larger than Pluto, and that it reflects a surprising 86 percent of light that strikes its surface. Pluto, by comparison, reflects only about 60 percent of light.

The announcement of an object larger than Pluto led to great excitement in the planetary community, and NASA even held a press conference announcing the discovery of the 10th planet!

Finally, the International Astronomical Union reopened the debate and took action to determine an answer. After a year or more of silence and indecision, in summer 2006 a new committee of the International Astronomical Union met in Paris and came up with the following new definition, which is now the accepted one:

Resolution 5: Definition of a Planet in the Solar System

Contemporary observations are changing our understanding of planetary systems, and it is important that our nomenclature for objects reflect our current understanding. This applies, in particular, to the designation "planets." The word "planet" originally described "wanderers" that were known only as moving lights in the sky. Recent discoveries lead us to create a new definition, which we can make using currently available scientific information.

The IAU therefore resolves that planets and other bodies, except satellites, in our Solar System be defined into three distinct categories in the following way:

(1) A planet[1] is a celestial body that
    (a) is in orbit around the Sun,

(continues)

(continued)

      (b) has sufficient mass for its self-gravity to overcome rigid body forces so that it assumes a hydrostatic equilibrium (nearly round) shape, and

      (c) has cleared the neighborhood around its orbit.

  (2) A "dwarf planet" is a celestial body that

      (a) is in orbit around the Sun,

      (b) has sufficient mass for its self-gravity to overcome rigid body forces so that it assumes a hydrostatic equilibrium (nearly round) shape[2],

      (c) has not cleared the neighborhood around its orbit, and

      (d) is not a satellite.

  (3) All other objects[3], except satellites, orbiting the Sun shall be referred to collectively as "Small Solar System Bodies."

———————

Notes

1. The eight planets are: Mercury, Venus, Earth, Mars, Jupiter, Saturn, Uranus, and Neptune.

2. An IAU process will be established to assign borderline objects into either a dwarf planet or other categorie(s).

3. These currently include most of the Solar System asteroids, most Trans-Neptunian Objects (TNOs), comets, and other small bodies.

As a corollary to Resolution 5, the IAU then issued Resolution 6, which declared that Pluto is no longer a planet.

Resolution 6: Pluto

The IAU further resolves:

Pluto is a "dwarf planet" by the above definition and is recognized as the prototype of a new category of Trans-Neptunian Objects[1].

———————

1. An IAU process will be established to select a name for this category.

After hearing that he was no longer a discoverer of the 10th planet, Mike Brown posted the following heartfelt text on his Web site:

> When I discovered it and realized that it was, indeed, bigger than Pluto, I immediately called my wife and excitedly told her "I found a planet!"

Right after the astronomical vote yesterday, I made the same phone call again. I had to tell her that the 10th planet was being buried alongside Pluto. Her voice dropped. Really? She said. Really. My wife was already mourning the little planet that we had gotten to know so well. I think her reaction was like that of the many Pluto fans out there who feel an emotional attachment to Pluto. See, to us Xena was more than just "the 10th planet." We had gotten to know her quite well over the past year. We knew about her tiny moon (Gabrielle, of course), her incredibly shiny surface, and her atmosphere frozen in a thin layer all around the globe. We had discussed her name, her orbit, and how many more like her might be out there.

Brown went on to say that he thought the IAU had done the right thing; after all, Pluto and 2003 UB313 are both significantly smaller than the other planets and do share their orbits with many other like objects.

After some internal debate, Brown, Trujillo, and Rabinowitz suggested Eris as the official name for 2003 UB313, dropping the informal nickname Xena, and the name was accepted by the International Astronomical Union on September 13, 2006. In Greek mythology, Eris is the goddess of warfare, and she stirs up jealousy to cause fighting among men. In the astronomical world, as Brown points out, Eris stirred up a great deal of trouble among the international astronomical community when the question of its proper designation led to a solar system with only eight planets. Eris's moon has received the official name Dysnomia, who in Greek mythology is Eris's daughter and the demon spirit of lawlessness.

Around the world people objected to the new designation of Pluto—the object has been entirely cemented into modern culture as a planet, and it has a large sentimental following. Bumper stickers were seen around the United States proclaiming, "Honk if Pluto is Still a Planet." Clearly, in the hearts of people, Pluto is still just as important as it ever was, despite the first-ever scientific definition of a planet, from which it is excluded. On the other hand, Pluto has entered an even more exclusive category, that of the dwarf planet. The IAU recognizes only five at the moment, Pluto, Eris, Ceres, Haumea (a Kuiper belt object with about one-third the mass of Pluto), and Makemake (another Kuiper belt object of similar mass to Haumea). However, about an additional 40 objects are under consideration for the designation.

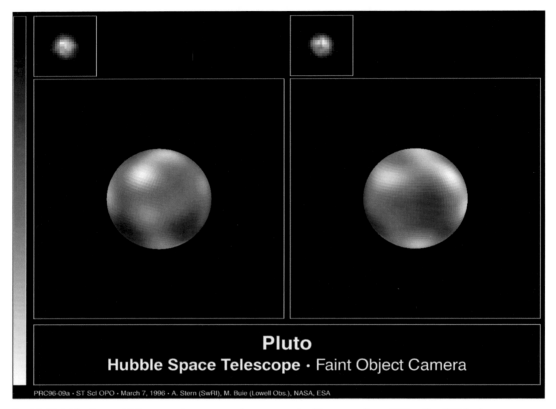

Pluto
Hubble Space Telescope · Faint Object Camera

PRC96-09a · ST ScI OPO · March 7, 1996 · A. Stern (SwRI), M. Buie (Lowell Obs.), NASA, ESA

In this Hubble Space Telescope *image, each square pixel is more than 100 miles (60 km) across, though this was the highest resolution image of Pluto ever made. At this resolution,* Hubble *discerns roughly 12 major regions where the surface is either bright or dark.* (Alan Stern [Southwest Research Institute], Marc Buie [Lowell Observatory], NASA and ESA)

(continued from page 122)
computer-processed images of each of Pluto's hemispheres. There are approximately 12 dark and light regions on the planet.

Pluto has a bright south pole and a dark equator, perhaps darkened by chemical reactions with the solar wind. The bright patches on Pluto's surface are thought to be methane (CH_4), nitrogen (N_2), and carbon monoxide (CO) ices, bright like snow on Earth. Water (H_2O) has also been detected in substantial amounts. The dark patches may be primordial organic matter, or they may be the result of millennia of cosmic ray bombardment turning simple hydrocarbons into more complex and darker molecules, or they may be something else entirely.

The color and density of Pluto are remarkably similar to those of Neptune's moon Triton, lending support to the theory that Triton was once a Kuiper belt object that was perturbed out of its orbit and captured by Neptune.

In June 1988 Jim Elliot and Leslie Young of the Southwest Research Institute detected an atmosphere on Pluto when it

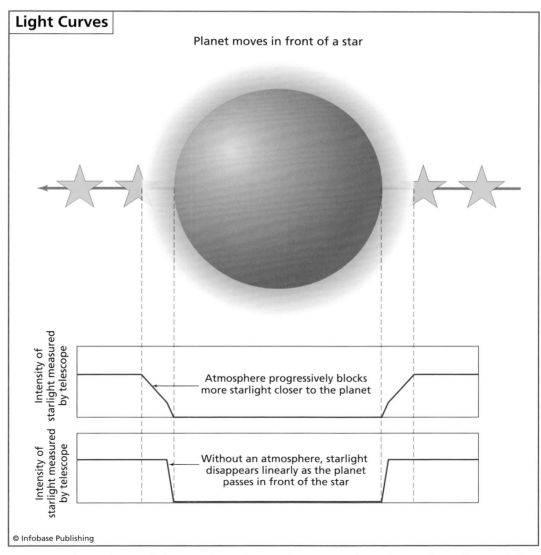

Light Curves

Planet moves in front of a star

Intensity of starlight measured by telescope

Atmosphere progressively blocks more starlight closer to the planet

Intensity of starlight measured by telescope

Without an atmosphere, starlight disappears linearly as the planet passes in front of the star

© Infobase Publishing

The bottom figure shows a light curve for a planet with no atmosphere; the top figure shows a light curve in the shape of Pluto's, showing the angle indicative of an atmosphere.

passed in front of a bright star, an event called a stellar occultation. During the stellar occultation the astronomers measured the intensity of the starlight. When a star passes behind a planet with no atmosphere, the intensity of starlight drops off quickly and smoothly as the planet passes in front of the star, and rises again smoothly as the star reemerges on the other side of the planet. This curve of light intensity versus time is called a light curve. A typical light curve for a planet with no atmosphere is shown on the left in the figure on page 129. In Pluto's case, the light curve dropped more slowly at first as the star passed behind the planet, and then dropped rapidly, as shown in the accompanying figure on the right. The best explanation to date for the shape of the light curve is that Pluto had a thin atmosphere at distances farther from the planet than the break in slope in the light curve, and that closer to the planet it had a haze layer with a steep temperature gradient.

Little is known about Pluto's atmosphere, but it is thought to consist of nitrogen (N_2), carbon monoxide (CO), and methane (CH_4). Methane is the only molecule definitively measured from spectroscopy, but nitrogen and carbon monoxide would be expected to sublimate into the atmosphere at the temperatures that allow methane to do so. Nitrogen is the most volatile gas on Pluto and also the most abundant (though water is thought to be the most abundant ice on the planet; at Pluto's temperatures ice is hard as rock is on Earth and will not exchange with the atmosphere). Based on thermodynamic considerations, the atmosphere should consist of 99 percent nitrogen, slightly less than 1 percent carbon monoxide, and about 1/10 of a percent methane. Pluto thus joins the Earth, Saturn's moon Titan, and Neptune's moon Triton as one of the few bodies in the solar system with an atmosphere dominated by nitrogen.

Pluto's atmosphere may only exist in any quantities near its perihelion, where temperatures on the planet are aided by additional solar heating and ices are more likely to sublime into gases. At perihelion the surface pressure is thought to be three

(continues on page 134)

THE *NEW HORIZONS* MISSION TO PLUTO

Pluto and its moon Charon reside in the Kuiper belt. Until 2006 Pluto was officially the ninth planet in the solar system, but change was in the air: In 2005 a team of scientists found a Kuiper belt object larger than Pluto. This object, Eris, is about 5 percent larger than Pluto. Immediately the question was raised: Is Eris a new planet? Or, in fact, is Pluto not a planet? The decision was made by the International Astronomical Union to demote Pluto to the status of a dwarf planet and to reduce our official number of planets to eight (see sidebar "How the Discovery of Eris Caused Pluto to Lose Its Status as a Planet" on page 123).

There are thought to be thousands of bodies in the Kuiper belt with diameters of at least 600 miles (1,000 km), 70,000 objects with diameters larger than 60 miles (100 km),

(continues)

An artist's drawing of the New Horizons *mission encountering Pluto* (NASA)

(continued)

and at least 450,000 bodies with diameters larger than 30 miles (50 km). The orbits of these bodies were first thought to merge into those of the Oort cloud, the immense cloud of comets that forms the outer solar system (described below with comets), but better observations show there is a clear outer edge to the Kuiper belt and a gap of about 11 AU before the Oort cloud begins. The Kuiper belt thus forms a contained population.

The first Kuiper belt body after Pluto was found in 1992 by David Jewitt at the Institute for Astronomy in Honolulu, using the Mauna Kea telescope. Its preliminary minor planet designation was 1992 QB_1, and it is about 150 miles (240 km) in diameter. The discoverers wished to name 1992 QB_1 Smiley, after a character from John le Carré's novels, but an asteroid had already claimed that name. Jewitt named a subsequent Kuiper belt discovery Karla, another le Carré character. A number of researchers are searching for Kuiper belt objects, and their searches have resulted in many discoveries. Kuiper belt object 28978 Ixion is at least 580 miles (930 km) across, about the same size as the largest inner-solar system asteroid, 1 Ceres. It appears to consist of a solid mixture of rock and ice.

Other large Kuiper belt objects (aside from Eris and Pluto) include Haumea (2003 EL61), Quaoar, Makemake (2005 FY9), Charon, Orcus, Ixion, and Varuna, all thought to be a least 600 miles (1,000 km) in diameter. Pluto itself has a diameter of approximately 1,450 miles (2,320 km). With the exception of Pluto, which will not likely get a more accurate size measurement until *New Horizons* visits it, all these bodies have multiple size estimates that have been produced by different research teams using different techniques. Estimates differing by hundred of kilometers can be found dating within a couple of years of one another. Thus, with the likely exception of Eris, stating unequivocally that one is larger than the other is a matter of argument.

Saturn's moon Phoebe is a captured asteroid, as evidenced by its highly inclined and retrograde orbit. The moon has prominent icy streaks inside its craters, indicating that it is an icy body with a mantle of dust and other material. Its high ice content may indicate that it originated far out in the solar system and, thus, may be the first object from the Kuiper belt to be observed closely by a space mission (not even Pluto has been visited).

New Horizons, launched on January 19, 2006, is the first of the New Frontiers program of medium-class planetary missions (the second, *Juno,* is planned to launch in 2011). *New Horizons* is planned to fly by Pluto and Charon. At the time of launch, Pluto was the last planet in the solar system never to have been visited by a spacecraft; eight months later, Pluto was demoted to a dwarf planet, but the importance of its visit is not diminished.

On its way to the Kuiper belt, where Pluto and Charon orbit, *New Horizons* received a gravity assist from Jupiter in February 2007. Passing at 51,000 km/hour (23 km/sec),

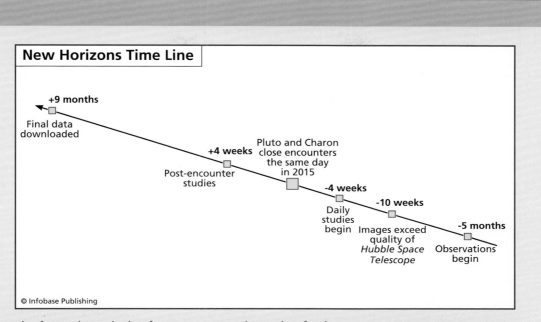

New Horizons Time Line

+9 months
Final data
downloaded

+4 weeks
Post-encounter
studies

Pluto and Charon
close encounters
the same day
in 2015

-4 weeks
Daily
studies
begin

-10 weeks
Images exceed
quality of
*Hubble Space
Telescope*

-5 months
Observations
begin

© Infobase Publishing

This figure shows the brief, intense, 14-month time line for the New Horizons *mission as it approaches, encounters, and passes Pluto and Charon in 2015.*

the craft was only 32 Jupiter radii away from the planet, about three times closer than the *Cassini* spacecraft. *New Horizons* will not reach Pluto and Charon until July 2015, and after it passes them, the craft will go on to a series of Kuiper belt encounters in the years 2016 to 2020. The whole mission will encompass 10 years and 3 billion miles.

When objects in the Kuiper belt are perturbed, they can fall into orbits that bring them into the inner solar system, where the solar wind excited jets of volatiles and turns them into comets. Because of the extreme distance to the Kuiper belt and the smallness of the objects, their total number is not known; even the mass of the Kuiper belt is not known within a multiple of two or more. Compositions are equally interesting—Kuiper belt objects are thought to be rich in organic material, the raw material from which life evolves.

Pluto's atmosphere is escaping to space like a comet, but on a planetary scale. Nothing like this exists anywhere else in the solar system. The primordial helium and hydrogen atmospheres of the young terrestrial planets were lost in this way. By studying Pluto's atmospheric escape, we can learn about the evolution of Earth's atmosphere. *New Horizons* will determine Pluto's atmospheric structure and composition and directly measure its escape rate for the first time.

(continues)

(continued)

The major mission objectives of *New Horizons* are the following:

1. Map the surface compositions of Pluto and Charon.
2. Characterize the geology and surface features of Pluto and Charon.
3. Characterize the atmosphere of Pluto and its escape rate.
4. Search for an atmosphere around Charon.
5. Map the surface temperatures on Pluto and Charon.
6. Search for rings and additional satellites around Pluto.
7. Conduct similar investigations of one or more Kuiper belt objects.

Though *New Horizons* will pass directly by Pluto and Charon without entering orbit, observations will still extend over a long period of time. The first meaningful observations of Pluto and Charon will begin five months before the closest encounter, and at 10 weeks out the images from the spacecraft will exceed the best *Hubble Space Telescope* images. Daily studies will begin at four weeks before the closest approach, and post-encounter studies will continue for a similar amount of time after the closest approach (see the figure on page 133). The extended mission is planned to include one to two encounters of Kuiper belt objects, ranging from about 25 to 55 miles (40 to 90 km) in diameter.

The exact Kuiper belt objects to be examined by *New Horizons* cannot yet be determined: There are so many years between now and when *New Horizons* will be near them that they are far in their orbits from the rendezvous locations. All of the objects that will be in the right place at the right time are now in line with the galactic plane from our point of view and, thus, cannot be detected in the brightness and mass of objects. This mission is in the unusual position of planning to visit objects that are not yet identified!

(continued from page 130)

to 160 microbars, about 1 millionth that of Earth's atmosphere. At aphelion, the atmosphere may well be completely frozen into surface ices. Pluto's atmosphere at perihelion extends to depths greater than Earth's atmosphere. Its atmosphere may even enclose Charon during the warmest periods. The atmosphere is thought to be actively escaping; Pluto is the only planet in the solar system actively losing its atmosphere now.

Pluto's atmosphere has been monitored since 1989, the planet's most recent perihelion. Throughout the 15 years

The science payload of *New Horizons* includes the following seven instruments:

1. **Ralph:** Visible and infrared imager/spectrometer; provides color, composition and temperature maps.
2. **Alice:** Ultraviolet imaging spectrometer; analyzes composition and structure of Pluto's atmosphere and looks for atmospheres around Charon and Kuiper belt objects.
3. **REX:** (Radio Science EXperiment) Measures atmospheric composition and temperature.
4. **LORRI:** (Long Range Reconnaissance Imager) telescopic camera; obtains encounter data at long distances, maps Pluto's farside and provides high resolution geologic data.
5. **SWAP:** (Solar Wind Around Pluto) Solar wind and plasma spectrometer; measures atmospheric escape rate and observes Pluto's interaction with solar wind.
6. **PEPSSI:** (Pluto Energetic Particle Spectrometer Science Investigation) Energetic particle spectrometer; measures the composition and density of plasma (ions) escaping from Pluto's atmosphere.
7. **SDC:** (Student Dust Counter) Built and operated by students; measures the space dust peppering *New Horizons* during its voyage across the solar system.

For more information about the *New Horizons* mission, and to see where the spacecraft is right now, visit the mission Web site at http://pluto.jhuapl.edu/index.php.

since Pluto's perihelion in 1989, its atmosphere has continued to expand. In this period the planet's surface atmospheric pressure has almost doubled. Though this may seem counterintuitive, since Pluto's atmosphere is expected to collapse back into an ice phase after perihelion, Pluto is experiencing normal seasons. Just as the hottest part of summer on Earth occurs well after the equinox, Pluto is experiencing further warming, up to its peak summer temperature. As Pluto continues in its orbit its temperatures will eventually fall and the atmosphere lessen, a state expected in about 10 years. Pluto reaches its next summer solstice in 2029 and its aphelion in 2114.

The surface temperature on Pluto varies between −390 and −328°F (−235 and −200°C). Two methods for measuring surface temperature, infrared blackbody radiation measurements and measurements of the shape of the nitrogen absorption bands, indicate different surface temperatures at the same time on the day side of Pluto. These day-side temperatures vary from −405 to −355°F (−243 to −215°C). Though the measurement techniques

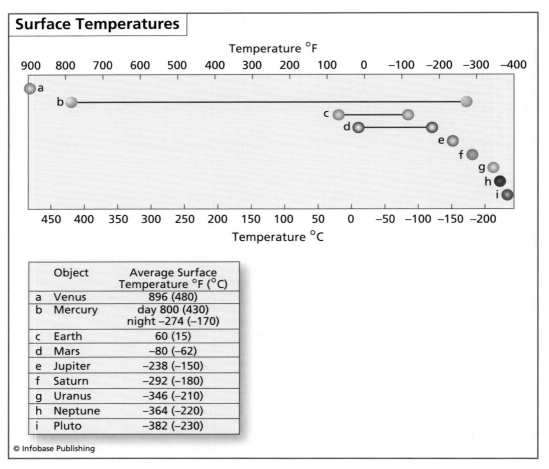

Surface Temperatures

Object		Average Surface Temperature °F (°C)
a	Venus	896 (480)
b	Mercury	day 800 (430) night −274 (−170)
c	Earth	60 (15)
d	Mars	−80 (−62)
e	Jupiter	−238 (−150)
f	Saturn	−292 (−180)
g	Uranus	−346 (−210)
h	Neptune	−364 (−220)
i	Pluto	−382 (−230)

© Infobase Publishing

The surface temperature ranges of each of the planets graphed here show that Mercury has by far the widest range of surface temperatures, though Venus has the hottest surface temperature, while Pluto, unsurprisingly, has the coldest.

have errors associated with them, the results indicate that the bright, icy areas of Pluto are measurably colder than the dark areas. This result is consistent with the colors of the regions, since dark material absorbs more heat than the reflective bright material. Pluto's surface is almost 1,300°F (700°C) colder than Venus's surface, as shown in the figure on page 136.

Pluto's atmosphere heats up rapidly with distance above the planet's surface. The gas giant planets all share the strange phenomenon of a hot uppermost atmosphere, but Pluto's hot atmosphere starts within a few kilometers of its surface. On Triton, for comparison, an equally high temperature is not reached before several hundreds of kilometers. Pluto's atmosphere reaches a high temperature of about −279°F (−173°C), as much as 126°F (70°C) hotter than the surface. Methane is particularly efficient at absorbing solar heat, and the high temperature may be explained by a gradient in methane abundance in the planet's outer atmosphere.

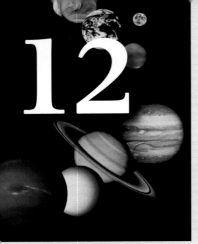

12

Charon: Pluto's Moon, or Its Companion Dwarf Planet?

In 1978, 48 years after the discovery of Pluto, the U.S. Naval Observatory scientists James Christy and Robert Harrington found Charon while trying to refine the orbital parameters of Pluto. Charon is only one-fifth as bright as Pluto, but their close mutual orbit had prevented the discovery of Charon's existence. Charon is named, in part, for the mythological figure who ferried the dead to Hades and is sometimes pronounced "KARE-en." But because Christy also named the moon in honor of his wife, Charlene, many planetary scientists pronounce the moon's name with a soft "ch" sound ("SHAHR-en").

Charon is the largest moon in relation to its planet in the solar system; the two form a double-dwarf planet system more than a planetary system with one moon. The most recent measurement for the semimajor axis of the Pluto-Charon orbital system is 11,580 miles (18,636 km), with an error of plus or minus five miles (8 km) on the measurement. Charon's orbit has an eccentricity of 0.0076, a slight but significant deviation from circularity. The tidal interactions between a planet and its moon normally have the effect of rapidly damping away any eccentricity (note that the eccentricities of the inner moons of Jupiter and Saturn are zero out to three decimal places). That

Charon's orbit has an eccentricity may indicate that the system was recently disturbed by interacting with a larger mass or by an impact with the last several million years.

The Jet Propulsion Laboratory has produced what may be the best estimate for Charon's radius: 364 miles plus or minus eight miles (586 km plus or minus 13 km). Other estimates for Charons' radius range from 360 to 391 miles (580 to 630 km). Though, as discussed above, the mass of the Pluto-Charon system is well known, its division between Pluto and Charon is not.

The consensus value seems to be that Charon contains about 11 percent of the total mass of the system, making its mass about 3.36×10^{21} pounds (0.15×10^{22} kg). Based on the best estimate for its radius and its mass, the density of Charon is between 95 and 114 pounds per cubic foot (1,500 and 1,800 kg/m^3). The ratio of masses is between 0.084 and 0.157, and a closer flyby mission is needed before their masses can be better calculated.

In truth, Charon does not strictly orbit Pluto: Charon's mass is close enough to Pluto's that they together orbit the center of mass of their combined system (in a system such as Mars and Phobos, Mars's mass is so much larger than Phobos's that their combined center of mass is well within Mars, and Mars's orbit is little affected by Phobos. For Pluto and Charon, their masses are similar enough that each body is affected by their mutual orbit. The orbital period of the two-body system has been exactly measured and found to be 6.387223 days, with a margin of error of plus or minus 0.000017 days.

Pluto and Charon are locked in a mutual synchronous orbit. The Moon always keeps the same face to the Earth, but the Earth turns independently. In the Pluto and Charon system, each body keeps the same face to the other at all times. Therefore, Charon is either always visible or never visible, depending on where one is on the surface of Pluto. The same is true for a viewer looking from Charon at Pluto: Pluto is either always or never visible, depending only on the viewer's location.

The mutual orbit of Charon and Pluto has an obliquity of about 120 degrees to the ecliptic plane and is thought to vary

An artist's interpretation of what the moons of Pluto might look like, viewed from Pluto's surface (NASA)

from about 103 to 128 degrees over a period of about 3 million years. This estimate is a result of numerical modeling of the torque of the Charon-Pluto system, not of observations.

Charon's surface lacks the strongly contrasting patches that cover Pluto. While Pluto's brightness varies by 30 percent over their six-day rotation, Charon's only varies by about 4 percent. Water ice seems to cover Charon's surface, and the moon's composition is thought to be similar to Saturn's icy moons, for example, Rhea. The water ice on Charon's surface is cold enough to attain the crystal structure of ice II. Unlike Pluto, Charon seems to have little or no nitrogen, carbon monoxide, or methane. Spectra from Charon are well fit by ammonia (NH_3) and ammonia hydrate ($NH_3°2H_2O$). Charon's albedo is about 0.4, in contrast to Pluto's albedo of about 0.7, further indicating that the two bodies have different compositions.

Charon and Pluto have different surfaces and are thought to be made of different materials. Though it was first suggested that Pluto and Charon began as a single body that was reformed into two by a giant impact (in the way the Earth and Moon were thought to be), their very different compositions make this unlikely. As discussed later, there are a large number of binary systems in the Kuiper belt, and Pluto and Charon simply seem to be the most massive. In this respect, as well as in their orbits and appearances, Pluto and Charon seem average denizens of the Kuiper belt.

On May 15, 2005, two additional moons of Pluto were seen in images from the *Hubble Space Telescope*. The two small bodies, named Nix and Hydra, orbit Pluto at about two and three times the distance of Charon. Nix and Hydra are thought to have diameters of less than about 88 miles (140 km) and 105 miles (170 km), respectively, but they are so small and distant that little is known about them. The top experts in imaging have trouble detecting these moons, so further information will have to come from the New Horizons mission.

The Rest of the Kuiper Belt Population

Neptune's orbit carries the planet from 29.8 to 30.3 AU from the Sun. Small bodies orbiting past Neptune are referred to as trans-Neptunian objects, and then are further subdivided into members of the Kuiper belt or the Oort cloud. Neptune marks the inner edge of the Kuiper belt. The Kuiper belt was originally thought to reach from 35 to 100 AU from the Sun, and then to merge into the Oort cloud of icy bodies. As study of the Kuiper belt has intensified and the orbits of more of its objects have been carefully calculated, there appears to be a gap between the edge of the Kuiper belt and the beginning of the comet-rich Oort cloud. The Kuiper belt begins around 30 AU and has a sharp outer edge at 49 AU. The reason for this gap is not understood; perhaps Kuiper belt bodies become fainter or smaller with distance and just cannot be seen as easily, or perhaps a sharp edge was formed by the disturbance of a passing planet or star, as unlikely as this event may be.

The Kuiper belt exists at extreme distances from the Sun. From the Kuiper belt the radius of the Sun appears 50 times smaller than it appears from Earth, which would make the Sun look more like a very bright star than something that dominates the day. Detecting objects at that distance from Earth is exceptionally difficult, and learning about their size and composition even more so. Only the most recent technology and largest telescopes allow the Kuiper belt to be explored.

The Kuiper belt had been postulated since 1943, but it remained a theory until 1992. Only the development of highly sensitive viewing instrument called a charge-coupled device has allowed astronomers to see the tiny bodies in the Kuiper belt. George Smith and Willard Boyle invented the charge-coupled device at Bell Laboratories in 1969, and once it was refined and put into mass production, it revolutionized cameras, fax machines, scanners, and, of course, telescopes. A charge-coupled device consists of many linked capacitors, which are electronic components that can store and transfer electrons. When a photon strikes the surface of the charge-coupled device, it can knock an electron off the atom in the surface it strikes. This electron is captured by the capacitors. In this case the capacitors are phosphorus-doped semiconductors, one for each pixel of the image. While photographic film records a paltry 2 percent of the light that strikes it, charge-coupled devices can record as much as 70 percent of incident light. Their extreme efficiency means that far dimmer objects can be detected. This sensitivity has made searching for small distant objects possible.

After a five-year search, David Jewitt and Jane Luu found the first Kuiper body other than Pluto in 1992 by examining series of photographs taken by the Mauna Kea telescope for moving bodies. The preliminary minor planet designation of this first body, about 150 miles (240 km) in diameter, was 1992 QB_1 (for more on these strange names, see the sidebar "Numbering and Naming Small Bodies" on page 146). The discoverers wished to name 1992 QB_1 Smiley, after a character from John le Carri's novels, but an asteroid had already claimed that name. The scientists named the next body they found, 1993 FW, Karla, also from le Carré's novels. By 2003 there were about 350 Kuiper belt objects known, and by 2004 more than 1,000 objects had been found. Now there are thought to be thousands of bodies in the Kuiper belt with diameters of at least 620 miles (1,000 km), about 70,000 with diameters larger than 60 miles (100 km), and at least 450,000 bodies with diameters larger than 30 miles (50 km).

Kuiper belt bodies are divided into three classes according to their orbits; classical (or *cubewano*), resonance (or plutino), or scattered disk. Classical Kuiper belt objects have orbits with low eccentricity and low inclinations, indicating that they formed from the solar nebula in place and have not been further perturbed. These objects are sometimes called cubewanos and include any large Kuiper belt object orbiting between about 41 AU and 48 AU but not controlled by orbital resonances with Neptune. The odd name is derived from 1992 QB_1, the first Kuiper belt object found. Subsequent objects were called "que-be-one-os," or cubewanos. There are more than 524 cubewanos known, including Varuna and Quaoar, described in more detail below.

Resonance Kuiper belt objects are protected from gravitational perturbation by integral ratios between their orbital periods and Neptune's. Like Pluto, many Kuiper belt bodies have orbits in periods of 3:2 with Neptune, which allows them to orbit without being disturbed by Neptune's gravity. Because they share their resonance with Pluto, this subclass of objects are called plutinos. There are about 150 plutinos and 22 other resonance objects known. Models indicate that only between 10 and 20 percent of Kuiper belt objects are plutinos, meaning there are likely more than 30,000 plutinos larger than 60 miles (100 km) in diameter. Though the Kuiper belt strictly begins at around 30 AU, the region between Neptune and about 42 AU is largely empty, with the exception of the plutinos, a large population of bodies that orbit at about 39 AU, a few bodies in the 4:3 resonance at 36.4 AU, and two objects, 1996 TR_{66} and 1997 SZ_{10}, which seem to be in a 2:1 resonance with Neptune at 47.8 AU. These bodies in the 2:1 resonance have perihelia close to Neptune's orbit. A few more Kuiper belt bodies have been found at the 5:3 resonance near 42 AU.

The large number of bodies in resonant orbits is another paradox of the Kuiper belt. How have so many bodies fallen into these orbits? Renu Malhotra, a scientist at the University of Arizona, suggests that interactions with the gas giants early in solar system formation can explain these highly populated orbits. Computer modeling efforts as early as the 1980s indicated that gas giant planets are likely to migrate outward in the solar system early in formation. The early solar system certainly had more material in the orbits of the planets than it does now, probably including multiple bodies as large as the Earth in the orbits of the outer planets. The gas giant planets would collide with these planetesimals and scatter them either inward, toward the Sun, or outward. Planetesimals scattered outward by Neptune, Uranus, and Saturn almost certainly returned inward through the force of the Sun's gravity to be scattered again, until at last they were scattered inward toward the Sun. During each collision and scattering event the giant planet in question has its orbit altered. Scattering a planetesimal inward toward the Sun drives the giant planet outward. Saturn, Uranus, and Neptune scattered more planetesimals inward than outward because those scattered outward returned to be scattered inward, and thus those three giant planets gradually migrated outward from the Sun. Jupiter, on the other hand, is massive enough that the planetesimals it scatters outward do not return under the Sun's gravity. Jupiter scattered slightly more planetesimals outward than inward, and so its orbit decayed slightly toward the Sun.

As Neptune, the most distant planet from the Sun, moved even further out, the locations of its resonant orbits moved outward ahead of it. These stable orbital positions thus could sweep up and capture small bodies that otherwise would never have encountered those resonant positions. Objects captured in Neptune's resonances also have their orbital eccentricities increased in a way predictable by theory. Pluto was ostensibly captured in this way, and to reach its current orbital eccentricity of 0.25, Neptune must have captured Pluto when Neptune was at 25 AU from the Sun and Pluto at 33 AU, in comparison to their current 30 and 39 AU respective distances from the Sun. Neptune continues to change the orbits of Kuiper belt bodies that are not in resonant orbits, and over the age of the solar system, Neptune is thought to have removed 40 percent of the Kuiper belt through gravitational interactions.

Before the kinds of careful and sophisticated modeling that made the preceding description possible, both classical and resonance Kuiper belt bodies were thought to be orbiting

(continues on page 150)

NUMBERING AND NAMING SMALL BODIES

The first small bodies to require special naming conventions were the asteroids. The first asteroid, 1 Ceres, was discovered in 1801, just a few decades after the discovery of Uranus. Ceres was the first object smaller than a planet that had been discovered orbiting the Sun. By approximately 1850 about 50 asteroids had been discovered, necessitating a system of temporary numbers called "provisional designations." Provisional designations are assigned to each new possible asteroid discovery and kept until the asteroid is confirmed as a new body. It was thought at the time that there would be no more than 26 new discoveries per half month, and so each half month of the year is assigned a letter: The first half of January is called A, the second half B, the first half of February C, and so on. Within each half month, new discoveries are given letter designations as well, with the first asteroid of each half month called A. For example, the first asteroid discovered in the second half of February of the year 2004 has the provisional designation 2004 DA (D for the second half of February, A for the first asteroid of that month).

Unfortunately, this system quickly became too constraining, as the rate of asteroid discoveries accelerated. By the 1890s, photographic film could be used to search for asteroids: If a camera's shutter is left open for some period of time, an asteroid moves fast enough to appear as a streak, while stars in the background are more stable. If more than 26 new asteroids are discovered in a half month, then the next asteroid gets the designation A_1, and the next B_1, and so on through the next alphabet, until it is used up and a third alphabet begins, with designation A_2, and so on. The last asteroid discovered in one especially productive half month was designated 1998 SL_{165}, meaning the namers had gone halfway through their 166th alphabet! This represents 4,136 objects discovered in that half month.

To be issued a final number and have its provisional designation taken away, a new asteroid's orbit must be determined closely, and also it must be confirmed that this is not a new sighting of a previously known object. The new asteroid must be observed at *opposition* (see figure on page 129) four times to make it an official part of the permanent record. In the year 2000, 40,607 minor planets were confirmed and numbered. That year was the high point, however; 36,459 of the discoveries in 2001 were numbered, 26,703 of those in 2002, 15,356 from 2003, and from 2008, fewer than 1,000. The history of minor planet discoveries by year is shown in figure on page 148. By September 2009 there were 219,018 confirmed and numbered minor planets in total, 161,322 unnumbered objects with fairly well deter-

Asteroids leave streaks on long-exposure photographs because they move more quickly than background stars. They are occasionally discovered by the streaks on photographs taken by the Hubble Space Telescope. (R. Evans & K. Stapelfeldt, JPL/WFPC2/HST/NASA)

mined orbits, and 79,931 unnumbered objects with poorly known orbits. There are thought to be millions of asteroids in the solar system, so searchers have a long way to go.

(continues)

(continued)

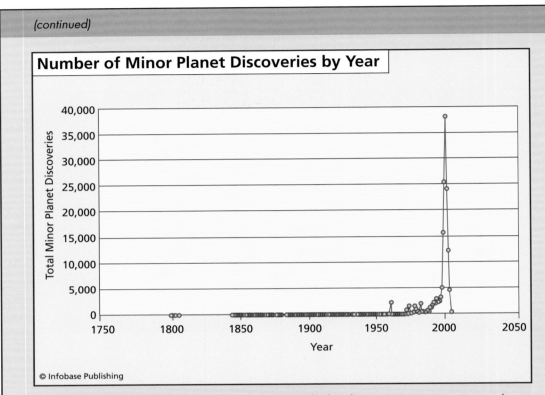

Number of Minor Planet Discoveries by Year

The number of minor planet discoveries by year peaked in the year 2000 as interest and technology converged on the problem; further discoveries become more difficult because of small size, dimness, and distance from the Earth.

Confirmed minor planets are numbered in the order of their confirmation. Thus, Ceres is number 1 and is formally notated 1 Ceres. Most (about 63 percent in 2005) still have names in addition to numbers, but only one has a name but no number: Hermes, whose provisional designation was 1937 UB.

Now, of course, scientists are concerned with more than just asteroids. Since the discovery of the first Kuiper belt object in 1992, the system of provisional designations, final numbers, and names has been expanded to include all small objects orbiting the Sun, including Kuiper belt objects and others found still farther out in the solar system. Final numbers are still assigned in order as the object's orbit is characterized and it is proven not to be a repeat discovery. The discoverer of the asteroid has a decade after the assignment of the final number to suggest a name, which must be approved by the 11-member Small Bodies Names Committee.

Opposition and Conjunction

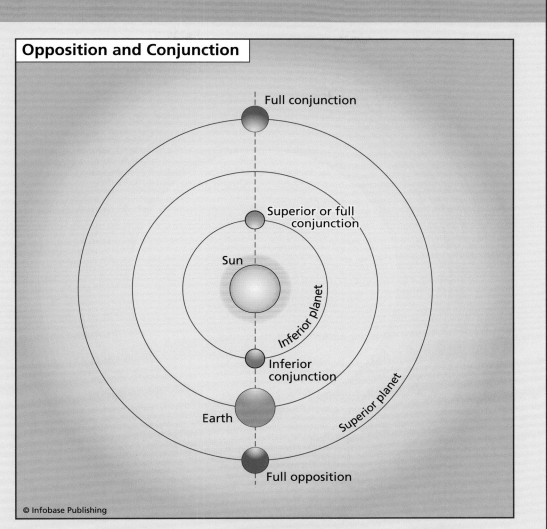

Full conjunction

Superior or full
conjunction

Sun

Inferior planet

Inferior
conjunction

Superior planet

Earth

Full opposition

© Infobase Publishing

Opposition and conjunction are the two cases when the Earth, the Sun, and the body in question form a line in space.

About the first 400 asteroids were named after figures from classical mythology, but since that time, many other categories of names have been used. While discoverers of asteroids can choose names from almost any category, other objects must have names from a particular subject. All the objects that share Pluto's orbit, for example, must be

(continues)

(continued)

named for underworld deities. Asteroids have been named after famous or accomplished people of all stripes, family members, committees, plants, and even machines. Asteroids with special numbers often get special names, as in 1000 Piazzi (the discoverer of the first asteroid), 2000 William Herschel, 3000 Leonardo da Vinci, 4000 Hipparchus, 5000 IAU (the International Astronomical Union), 6000 United Nations, 7000 Marie and Pierre Curie, 8000 Isaac Newton, and 9000 Hal (named for the computer, Hal 9000, in the 1968 movie *2001: A Space Odyssey*). The asteroid 6765 Fibonacci is named because 6765 is a number in the Fibonacci sequence, a mathematical sequence discovered and investigated by Leonardo Fibonacci.

(continued from page 145)

at the points of their formation in the solar system, largely undisturbed since the beginning of the solar system. Unlike the randomly oriented orbits of the Oort cloud objects, Kuiper belt objects have orbits closer to the ecliptic plane. Most of the major planets' orbital planes form angles of less than a few degrees with the ecliptic plane. Neptune's orbital plane lies at a little less than two degrees from the ecliptic plane, but Pluto's orbit is the most highly inclined of the planets, at just over 17 degrees. Pluto's high inclination also marks it as a typical Kuiper belt object. A number of Kuiper belt objects have inclinations larger than 25 degrees, and the highest inclination is $46.6°$ for 2004 XR_{190}, which orbits at about 57 AU. Because the Kuiper belt objects have nonzero orbital inclinations, the Kuiper belt itself has a thickness, in contrast to the planets out to Neptune, which almost define a plane. The Kuiper belt's thickness is about 10 degrees. In addition to the range of orbital inclinations for classical and resonance Kuiper belt bodies, their orbits have eccentricities up to 0.4. These ranges indicate that these objects must have been disturbed from their original orbits, which should have been almost circular and close to the ecliptic plane. They may have been disturbed by larger bodies that existed in the early Kuiper belt but have now been destroyed through collisions, or the disturbance may have been caused by Neptune moving its orbit outward, but the dynamics of this process are not well construed.

The third class, scattered disk Kuiper belt objects, has large, eccentric orbits, perhaps created by gravitational interactions with the giant planets. The Kuiper belt object 1996 TL_{66} is a good example of this class, with an orbital eccentricity of 0.59 that carries it to 130 AU at aphelion. There are thought to be as many as 10,000 scattered Kuiper belt objects. The figure below shows the nearly circular orbits of the plutinos and the widely eccentric orbits of a few of the scattered objects.

The inner planets are thought to have accreted from collisions between smaller bodies that finally led to one large body in a given orbit. The Kuiper belt bodies are clearly not in this state, having survived to this point in the evolution of the solar system as a population of tens of thousands of small

This sketch of the range of orbits of the plutinos and a few selected scattered Kuiper belt objects shows not only their immense distance from the Sun (compare to Jupiter's orbit) but also the wide range of orbital shapes and sizes in the Kuiper belt.

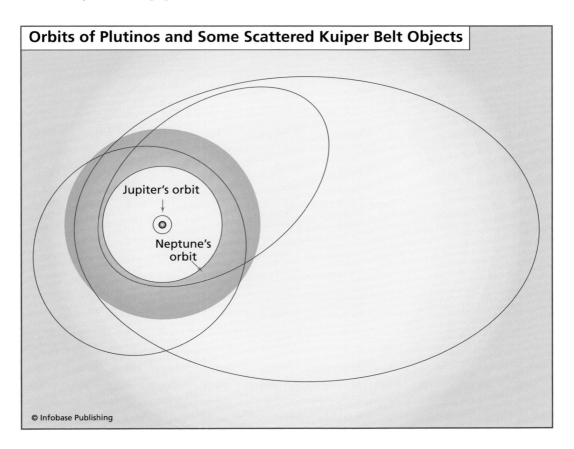

Orbits of Plutinos and Some Scattered Kuiper Belt Objects

Jupiter's orbit

Neptune's orbit

© Infobase Publishing

This artist's concept image shows the moment of impact of the mission Deep Impact, when the craft formed a man-made crater in the comet. (NASA/JPL/UMD)

bodies. Collisions among current Kuiper belt bodies are infrequent enough that they cannot be building now into a larger body, and in fact, the high eccentricities and obliquities of their orbits make their collisions more likely to break bodies up into smaller pieces than to join them together into larger bodies. Earlier in solar system history, before Neptune had sufficiently perturbed the Kuiper belt bodies into their current range of orbits, collisions may have led to accretion and allowed the building of the larger bodies now seen.

Though scientists believe short-period comets originate in the Kuiper belt, they do not know which population of Kuiper belt objects is most likely to provide the comets. Neptune, however, is thought to be the main provider of the gravitational perturbations needed to throw a Kuiper belt object into the inner solar system. Kuiper belt objects that are not in stable resonant orbits are thought to experience a close encounter with Neptune on the average of once every few tens of millions of years. This low but constant probability of Neptune encounters provides the inner solar system with a small, constant supply of short-term comets. A Kuiper belt object that comes close to Neptune has about a one-third probability of being moved into a short-period comet orbit, and it will otherwise continue in a new Kuiper belt orbit, be ejected from the solar system, or collide with a planet.

In general a Kuiper belt object should have enough volatiles to continue producing a tail as a comet for about 10,000 years, and short-period comets have average lifetimes of about 100,000 years before colliding with a larger body or being expelled from the solar system by gravitational forces. Some orbiting objects, then, should be intermediate between comets and asteroids as they lose their volatiles. Other bodies should be in the process of moving from the Kuiper belt into short-period comet orbits. The Centaurs, bodies that orbit neat Saturn and Jupiter, are thought to be these bodies in transition. The Centaur 2060 Chiron has a cometary coma (a cloud of gas

NASA's Spitzer Space Telescope *took this image, showing the comet Encke orbiting along its pebbly debris trail (long diagonal line). Encke has left debris over its entire orbital path. When the Earth passes through Encke's debris trail, as it does every October, the debris creates the Taurid meteor shower. Jets of material can also be seen shooting away from the comet.* (NASA/JPL-Caltech/University of Minnesota)

and dust encircling the solid body), supporting the theory that it originated in the Kuiper belt and may be perturbed into a cometary orbit.

The albedo of a body, the percent of sunlight it reflects, allows the body's size to be calculated. If the albedo is known and the sunlight intensity reflecting from the body is measured, then its size can be calculated. Most Kuiper belt objects are thought to have an albedo of about 4 percent, making them about as dark as charcoal. Based on a 4 percent albedo, the majority of Kuiper belt objects detected so far are judged to be typically about 60 miles (100 km) in radius.

The value of 4 percent albedo is a guess, however, and albedo is certain to differ among the objects. Albedo can be calculated by comparing the amount of sunlight reflected from a body with the amount of infrared radiation it emits. The infrared radiation is created by absorbed sunlight; it and the reflected sunlight should make up the total of sunlight striking the body. With a few assumptions about the material the body consists of (which controls how the sunlight heats the body and therefore its infrared emissions), the albedo of the body can be calculated.

The Kuiper belt body Varuna, for example, is thought to have an albedo of about 7 percent. For the same intensity of reflected light, a body with 7 percent albedo would be estimated to be smaller than a body with 4 percent albedo, which needs more surface area to reflect the same amount of sunlight. Pluto is exceptionally bright, with an albedo of about 60 percent. Its high albedo is thought to be created by constant cycles of ice sublimation and subsequent crystallization of fresh ice on the planet's surface as its atmosphere rises and freezes with its seasons. Young ice is highly reflective. A similar process may explain how dark Kuiper belt objects become bright short-period comets if they are perturbed into the inner solar system: Solar heating burns off the body's dark, weathered rind, and fresh icy and gassy material emerge from the interior.

Because of the extreme faintness of Kuiper belt objects, spectra are difficult to obtain. Most inferences about Kuiper belt objects have come from broadband colors, that is, the colors that the objects appear to the eye. The objects have a wide range of colors that may indicate either a range of compositions or that collisions have created fresh surfaces on some, while others retain surfaces that have been weathered by millennia in space. Many of them are exceptionally red. When solar system bodies are referred to as *reddened,* their color appears reddish to the unaided eye, but more important, the bodies have increased albedo (reflectance) at low wavelengths (the "red" end of the spectrum).

The unusual redness of Kuiper belt objects is not well understood, though laboratory tests show that the red spectrum of 5145 Pholus, a Centaur that orbits in the vicinity of Saturn and Uranus, can be reproduced by irradiating ices that contain nitrogen and methane. Nitrogen and methane are known constituents for Kuiper belt objects, and in fact, Centaurs are likely to have been Kuiper belt objects that were perturbed from their original orbits. The complex organic molecules produced by irradiating simple organic molecules are called tholins (named by the famed Cornell University astronomer Carl Sagan after the Greek word *tholos,* meaning "mud"). Though tholins are definitely red, they unfortunately have no specific spectral absorptions and so cannot be definitively recognized remotely on Kuiper belt objects. Irradiation is certainly a significant factor in the development of surfaces in Kuiper belt objects, which have no inherent magnetic fields to shield them from cosmic rays and solar wind. When these ices are struck by high-energy particles, they lose the relatively light element hydrogen and form additional carbon-carbon bonds. Cosmic rays are highly energetic and can penetrate ices to a distance of several yards, but weak solar radiation can penetrate only a few microns.

The red color may also be caused by weathered silicate minerals, since spectral analysis of a few Kuiper belt bodies indicates that the mineral *olivine* may be present. Kuiper belt objects are also proven by spectral analysis to contain water ice, though it would be hard as rock at the temperatures in the Kuiper belt. These objects should also contain a large proportion of dust in addition to their ices. Comets near the Sun eject more dust than gas, and their ratios may indicate something about the bulk compositions of Kuiper belt objects. The non-ice dust in the outer solar system is likely to be rich in radiogenic elements such as potassium, thorium, and uranium. These elements emit heat when they decay, and this heat may alter the compositions and structures of the Kuiper belt bodies.

Small Kuiper belt bodies can conduct their internal heat into space efficiently because of their small radii: Heat has a shorter distance to travel before escaping into space. Larger bodies can retain internal heat over the age of the solar system (as the Earth has done) and so may experience internal alterations because of this heat. They may partially differentiate, and internal gases may sublimate and move in response to the heat, and these gases may leave the body or even lead to cryovolcanism. Heat is transferred through a body in a characteristic time, T, that depends on the thermal diffusivity of the material, κ. Thermal diffusivity has units of area per second and measures the ease with which heat moves through the material in question. The thermal diffusivity of ice is 10^{-5} square feet per second (10^{-6} m^2/sec). The radius through which heat moves with time is given as approximately

$$r = \sqrt{\kappa T}.$$

Over the age of the solar system (T = 4.56 billion years, or 1.43 $\times$ 10^{17} seconds), heat can move from the center to the surface of a sphere with radius *(r)* equal to 310 miles (500 km). There are thousands of Kuiper belt bodies larger than this critical radius, and all these larger bodies are expected to show some internal changes from radiogenic heat. In particular, radiogenic heat should be sufficient to sublimate carbon monoxide and nitrogen, which might move toward the surface and then freeze again as they reach cooler, shallower depths. The Kuiper belt objects may thus acquire a layered structure.

It is a strange coincidence that Pluto and Charon are two of the largest Kuiper belt objects (see table on page 158), and they happen to be in a binary system with each other. At least a few percent of large Kuiper belt bodies are binaries, with orbits that have radii hundreds of times the radii of the bodies involved. There are 10 binaries known now, including 1997

CQ_{29} and 2000 CF_{105}, systems orbiting about 3,800 miles (6,100 km) and 6,300 miles (10,200 km) apart, respectively. The distance between these bodies is about 1/50 of the distance from the Earth to the Moon.

Binaries can be created by collision, by tidal capture, or by three-body interaction, in which three bodies pull toward each other and swing around each other, and then one body escapes, leaving the other two in a binary. The only way to make the binaries with objects of very similar mass, at great radii, is the third method. Peter Goldriech, at the Institute for Advanced Study at Princeton believes that almost all Kuiper belt objects are binaries. Observers have trouble discriminating them from each other: The binaries look like single objects from the distance of Earth.

Despite their surface irradiation, Kuiper belt objects are probably the least-altered objects in the solar system. Computer simulations by Matt Holman at the Harvard-Smithsonian Center for Astrophysics, Jack Wisdom at the Massachusetts Institute of Technology, and their colleagues show that Kuiper belt bodies can survive for the age of the solar system in a selection of their current, stable orbits. This study implies that Kuiper belt bodies are the remnants of the solar nebula that have stayed frozen and unaltered in the outer solar system for the last 4.56 billion years. Observations of the Kuiper belt, then, are literally observations of the original solar nebula itself. The Kuiper belt is thus a critical region for understanding solar system development, and it is a region targeted with ongoing and active research. Since 1992 a number of large bodies have been found in the Kuiper belt. Some of the largest Kuiper belt bodies are listed in the table, opposite.

David Jewitt and his colleagues discovered one of the first very large Kuiper belt bodies, 20000 Varuna, in November 2000 using the Spacewatch telescope in Arizona. Originally known as 2000 WR_{106}, Varuna is thought to be about 560 miles (900 km) in diameter. Its diameter is therefore less than half of Pluto's. It is about the same size as 1 Ceres, the largest asteroid, which is 577 by 596 miles (930 by 960 km). Until Varuna was found, researchers thought that all Kuiper belt objects might have albedos of about 4 percent (with the exception of

SELECTED KUIPER BELT AND OORT CLOUD OBJECTS IN APPROXIMATE SIZE ORDER

Object	Diameter [miles (km)]	Discovery and comments
Eris (2003 UB$_{313}$)	1,500±60 (2,400±100)	Mike Brown, Chad Trujillo, and David Rabinowitz, 2003; the discovery of Eris started the dispute in the International Astronomical Union that resulted in the demotion of Pluto from planet status; Eris is officially a dwarf planet, and it has a moon called Dysnomia
Pluto	1,450 (2,320)	Clyde Tombaugh, 1930; Pluto is now called a dwarf planet
Haumea (2003 EL$_{61}$)	1,200 (1,960) in its longest dimension	Mike Brown, Chad Trujillo, and David Rabinowitz, 2003; has two moons, Hi'iaka and Namaka; football-shaped
Makemake (2005 FY$_9$)	~1,070 (~1,730)	Mike Brown, Chad Trujillo, and David Rabinowitz, 2005
90377 Sedna (2003 VB$_{12}$)	~1,000 (~1,600 km)	Mike Brown; may be the first object found in the Oort cloud
50000 Quaoar (2002 LM$_{60}$)	750±120 (1,200±200)	Chad Trujillo and Mike Brown; cubewano; has one moon
Charon	728 (1,172)	James Christy and Robert Harrington, 1978
Orcus (2004 DW)	~590 (~940)	Mike Brown and Chad Trujillo; plutino; has one moon
28978 Ixion (2001 KX$_{76}$)	~580 (~930)	Lawrence Wasserman and colleagues at the Deep Ecliptic Survey; plutino
20000 Varuna (2000 WR$_{106}$)	560±90 (900±140)	Robert S. MacMillan, Spacewatch project; cubewano

Object	Diameter [miles (km)]	Discovery and comments
55565 (2002 AW$_{197}$)	555±75 (890±120)	Chad Trujillo and Mike Brown; cubewano
55636 (2002 TX$_{300}$)	~440 (~700)	Jet Propulsion Laboratory NEAT program; cubewano
1999 CL$_{119}$	~270 (~430)	Most distant object in the Kuiper belt (perihelion 46.6 AU, no other perihelion farther); cubewano
15760 (1992 QB$_1$)	150 (240)	David Jewitt and Jane Luu; the definitive cubewano and first discovered Kuiper belt object

(Note: Diameters marked as approximate (~) may have errors in the hundreds of kilometers; measurements are difficult, and new papers with new size measurement using new techniques are being published constantly, so a wide range of estimates is in the literature.)

brilliantly bright Pluto). Varuna seems to have an albedo of about 7 percent. Its brightness energized the science community, since the brighter the objects are, the easier they are to find, and so more searches for Kuiper belt objects might be successful. Initially, Varuna was thought to be as large as Charon, but with more refined calculations of its albedo the discovery team announced that it was in fact significantly smaller. Still, it was and is one of the larger objects yet discovered. Since Varuna, there have been at least four Kuiper belt objects discovered that are larger.

In July 2001 Robert Millis and his colleagues at the Massachusetts Institute of Technology, Lowell Observatory, and the Large Binocular Telescope Observatory in Arizona discovered 28978 Ixion (originally 2001 KX$_{76}$), a large Kuiper belt object (this was largely the same team that discovered Uranus's rings in 1977). Along with Varuna, Ixion is about the same size as 1 Ceres (see figure on page 161). Ixion's size was highly uncertain when it was discovered because the telescope through which it was discovered had neither the high resolution required to measure size directly (though the *Hubble Space Telescope* can do this) nor the ability to measure infrared radiation, which is related to size. At the time of discovery, the team thought Ixion was at least 750 miles (1,200 km) in diameter. This estimated size made a big media splash, since at the time it would have been the largest Kuiper belt object after Pluto and in fact larger than Charon itself. Further study shows Ixion to be only about 580 miles (930 km) in diameter, and in the intervening years a number of larger Kuiper belt bodies have been discovered, including at least two that really may be larger than Charon.

In June 2002 California Institute of Technology scientists Chad Trujillo (now at the Gemini Observatory) and Mike Brown saw for the first time a Kuiper belt body with the preliminary name 2002 LM_{60}, later named 50000 Quaoar (pronounced "KWAH-o-wahr"). Quaoar is named for the god found in the creation stories of the Tongva tribe, early inhabitants of what is now southern California. Quaoar lies at about 42 AU from the Sun. Its orbit takes about 285 Earth years and is almost circular, with an eccentricity of only 0.04 and an inclination of about eight degrees. Pluto's orbital eccentricity is about six times larger than that of Quaoar's, and Pluto's inclination is about twice Quaoar's. Because Quaoar is so bright, within a month of discovery they were able to trace its position back two decades in previously taken telescope images. Quaoar has a diameter of 750 miles (1,200 km), about half the size of Pluto. Quaoar was at the time of its discovery the largest solar system body found since Pluto itself.

In February 2004 Brown, Trujillo, and their colleague David Rabinowitz from Yale University had a new announcement about the then-largest-known Kuiper belt object, having found a new object designated 2004 DW that is still larger than Quaoar. Based on its current distance of about 48 AU from the Sun, its brightness, and its presumed albedo, 2004 DW, now known as Orcus, has been estimated to be about 870 to 990 miles (1,400 to 1,600 km) in diameter, or more than half the size of Pluto. As has happened with many of these discoveries, its size has been reestimated and now stands at about 580 miles (930 km). As with many other found objects, once they are identified they can then be found in photographs

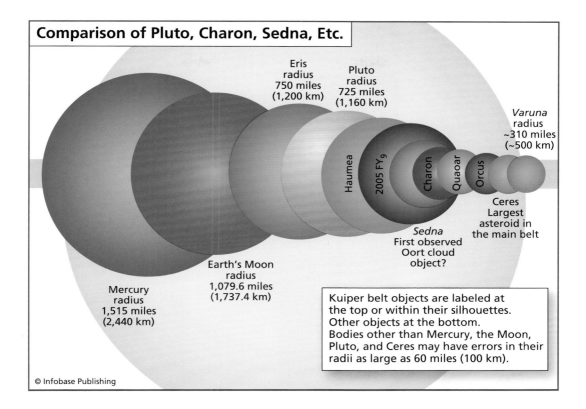

Comparison of Pluto, Charon, Sedna, Etc.

Eris
radius
750 miles
(1,200 km)

Pluto
radius
725 miles
(1,160 km)

Varuna
radius
~310 miles
(~500 km)

Haumea

2005 FY₉

Charon

Quaoar

Orcus

Ceres
Largest
asteroid in
the main belt

Sedna
First observed
Oort cloud
object?

Earth's Moon
radius
1,079.6 miles
(1,737.4 km)

Mercury
radius
1,515 miles
(2,440 km)

Kuiper belt objects are labeled at
the top or within their silhouettes.
Other objects at the bottom.
Bodies other than Mercury, the Moon,
Pluto, and Ceres may have errors in their
radii as large as 60 miles (100 km).

© Infobase Publishing

from sky surveys in the past. The object Orcus has been found in a First Palomar Sky Survey photograph of November 23, 1954, and in a November 8, 1951, photograph from the Siding Spring Observatory in Australia. It appears to be a plutino with an orbit that carries it from 30.9 and 48.1 AU with an orbital inclination of about 20.6 degrees. It requires 248 years to complete its orbit. It reached aphelion in 1989 and will reach perihelion in 2113.

In July 2005 the hardworking team of Brown, Trujillo, and Rabinowitz made their historic announcement: They had confirmed the existence of a Kuiper belt body larger than Pluto. The scientists had searched for outer solar-system objects using the Oschin Telescope at Palomar, California. It has a mirror diameter of 3.9 feet (1.2 m), which is large compared to amateur telescopes (typically ranging from 0.3 to one foot [0.1 to 0.3 m] in diameter), but small compared to most professional

A comparison of the sizes of Mercury, the Moon, Pluto, and a series of minor planets shows that all known Kuiper and asteroid belt objects are smaller than Pluto, but even Pluto is smaller than Earth's Moon.

Voyager 2 was launched August 20, 1977, using a Titan-Centaur rocket, 16 days before Voyager 1. Their different flight trajectories caused Voyager 2 to arrive at Jupiter four months later than Voyager 1, thus explaining their numbering. After completing the tour of the outer planets in 1989, the Voyager spacecraft began exploring interstellar space. (NASA/GRIN)

TITAN/CENTAUR COMPLEX 41

telescopes (3.28 to 32.8 feet [1 to 10 m] in diameter). The scientists examined a sequence of high-resolution telescope images of the same region in space, looking for objects that moved relative to the starry background (the stars are moving, also, but they are so far away relative to Kuiper belt objects that their movement is undetectable over small times). This new body, temporarily named 2003 UB_{313}, was first seen in 2003, but only recognized to be moving relative to background stars in January 2005. Now it is known to orbit at 44 degrees to the ecliptic, with a perihelion of 36 AU and an aphelion of 97 AU. It is thought to have a diameter of about 1,500 miles (2,400 km), making it the largest solar system body discovered since the discovery of Neptune in 1846. Discovery of this object, now known as Eris, caused the International Astronomical Union to redefine solar system bodies and reclassify Pluto as a dwarf planet (see sidebar on page 123). Trujillo and Brown have said they expect to find five to 10 more large objects, and perhaps some larger than Pluto.

Though Jewitt and others think Kuiper belt objects as large as Mars may remain undiscovered in the outer Kuiper belt, it is unlikely that any Kuiper belt object exists of a size comparable to the larger planets. Any large body in the outer solar system

The density of material in the solar system decreases with distance from the Sun.

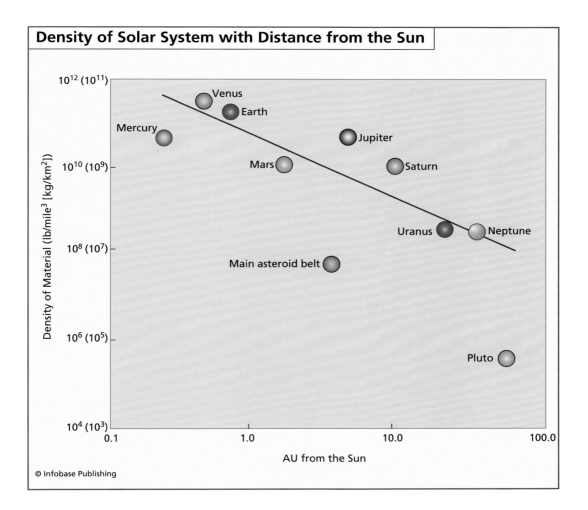

Density of Solar System with Distance from the Sun

Venus
Earth
Mercury
Jupiter
Mars
Saturn
Uranus
Neptune
Main asteroid belt
Pluto

Density of Material (lb/mile^3 [kg/km^2])

10^{12} (10^{11})

10^{10} (10^9)

10^8 (10^7)

10^6 (10^5)

10^4 (10^3)

0.1   1.0   10.0   100.0

AU from the Sun

© Infobase Publishing

would have perturbed the paths of the spacecraft *Voyager 1*, *Voyager 2*, and *Pioneer 10* as they passed the solar system, and should further influence the orbit of comet Halley. To have avoided changing the orbits of these objects, any remaining large Kuiper belt objects are thought to be less than five Earth masses, which in its turn is far larger than the estimated total mass of the Kuiper belt. According to Kuiper's original calculations as well as computer simulations of losses to the inner solar system and gravitational expulsions into outer space, the Kuiper belt's original mass in the beginning of the solar system was the equivalent of about 30 Earth masses. Now it is thought to be 0.2 Earth masses, or about 100 times the mass of the asteroid belt.

The figure on page 163 demonstrates the loss of mass from the inner solar system, particularly from the asteroid belt. The mass of matter at a given orbit divided by the area of the orbit is plotted on the vertical axis as a measure of density of material existing at that distance from the Sun. The horizontal axis measures distance from the Sun in AU. Theories and models of the solar nebula clearly indicate that densities in the solar nebula should smoothly decrease from the Sun outward, but this graph clearly shows that the decrease in density is not smooth. Could the excess mass needed to smooth this graph exist in the outer solar system as more moderately large bodies, like Pluto? There may be bodies even as large as Mars remaining undiscovered in the outer solar system.

The Nice Model for Kuiper Belt Formation

Over the years since 1992, more than 1,000 Kuiper belt bodies have been discovered and their orbits described. Tens of thousands of additional bodies likely await discovery, but the thousand that are now known are thought to form a representative sample.

Studying this population along with the outer planets has revealed some strange properties: First, judging from the density of material in the solar system with distance from the Sun, Uranus and Neptune are too large to have formed at their current orbits. There would have been insufficient material to accrete into such large planets. Second, the Kuiper belt has too little mass. At its distance, and over its range of orbits, more material would be expected. Since detecting the largest bodies is easiest, the tens of thousands of additional bodies yet to be discovered cannot add sufficient mass to make the Kuiper belt as large as it should be. All of the Kuiper belt objects together are thought to total just 0.01 to 0.1 times the mass of the Earth. For the larger bodies in the Kuiper belt to have grown through accretion to their present size, more original mass (more small bodies) are thought to have been required early in the solar system. Modeling efforts indicate that 100 to 1,000 times more mass had to have been present in Kuiper belt orbits at the beginning of the solar system.

Another set of questions about the Kuiper belt involves the orbits of the bodies. If the orbital periods of two bodies form an integer ratio, such as the 2:3 orbits of Pluto and Neptune (Pluto orbits the Sun twice for every three times Neptune orbits), they are said to be in resonance. Pluto and Neptune interact gravitationally in a regular way as they pass each other in this pattern of orbits.

Several kinds of resonances exist. Resonance created by orbital period, as in the example above, is called mean motion resonance. Resonance can be created not just with orbital period but also with any combination of orbital parameters. A resonance created by precession (regular change in the angle) of orbital perihelion, or any other angular measure of an orbit, is called secular resonance. In this case, the two orbits involved precess at the same rate.

These and other kinds of resonances can lead to stability, as the 2:3 Pluto to Neptune resonance does, since it keeps the bodies sufficiently far away from each other so that Pluto is not pushed out of its orbit by Neptune's gravity. Every time the small body completes one orbit, Neptune will be one-half-orbit away. Resonances more often lead to instability and are a significant way that small bodies are thrown out of the solar system, or into the Sun. The Kirkwood gaps in the asteroid belt are an example. In the main asteroid belt, there are gaps at semimajor orbital axes of 2.06, about 2.82, 2.95, and 3.27 AU, where few or no asteroids exist. They are named for the astronomer Daniel Kirkwood, who first noticed them in 1857 and correctly explained that those orbital sizes correspond to mean motion resonances with Jupiter of 3:1, 5:2, 7:3, and 2:1. When small bodies have orbits in those resonances, the regular gravitational interaction with Jupiter destabilizes their orbits, rather than stabilizing them, and so those orbits are gradually depopulated over time.

The ability of resonances to either stabilize or destabilize small bodies creates another question about the Kuiper belt. Orbits within the Kuiper belt that have strong stabilizing mean motion resonances have excess populations of small bodies in them. The 2:3 Pluto to Neptune resonance has about 200 known objects in it along with Pluto. Together these objects are known as Plutinos. Other significant stabilizing mean motion resonances in the Kuiper belt are the 2:5, 3:4, 3:5, and 4:7, each referring first to the small body in the Kuiper belt and second to Neptune.

The inner and outer edges of the Kuiper belt are also strangely sharp. The inner edge of the Kuiper belt begins abruptly around 39 AU, with the 2:3 Plutino resonance. Inside that orbit fewer than about 100 objects have been detected. Similarly, at 50 AU there is a sudden decrease in the number of objects. These observations together show a highly structured Kuiper belt. With relatively few exceptions, the belt begins at the 2:3 mean motion resonance with Neptune, which more than 200 bodies share with Pluto. The classical Kuiper belt extends from there to about 48 AU, punctuated by the populous orbits of 3:5, 4:7, and 1:2 mean motion resonances with Neptune.

This structure implies that the orbits of the giant planets migrated, disturbing the orbits of the many Kuiper belt objects. Though scientific thought has developed far beyond the ideas that all objects orbit in fixed spheres—meteorites are accepted to be pieces of early solar system material still migrating among and interacting with the planets, for example—only in recent years has the community begun to completely incorporate ideas of large planets changing their orbits.

Beginning in the 1990s with a series of papers by Renu Malhotra from the University of Arizona, attention began to focus on the likely migrations of Neptune and Uranus. All the giant planets are far more likely to have accreted to their present sizes closer to the Sun, where there was more mass in the planetary nebula. They then had to migrate outward to their present orbits. The implications of their migration and effects on other bodies were eventually developed into a detailed model for early evolution of solar system dynamics by a group of four people: Kleomenis Tsiganis from the Aristotle University of Thessaloniki, Greece; Rodney Gomes of the Observatório National, Rio de Janeiro, Brazil; Alessandro

Morbidelli of the Observatoire de la Côte d'Azur, Nice, France; and Harold Levison from the Southwest Research Institute in Boulder, Colorado.

Working together over a period of time at the Observatoire de la Côte d'Azur in Nice, the team developed a model, first published in the journal *Nature* in 2005. Theirs is the best and most comprehensive explanation for the observations made today of Kuiper belt structure, mass of the giant planets, and even the Late Heavy Bombardment, the period of increased meteorite impacts around 4.0 billion years ago that created the giant, basalt-filled craters on the Moon. In honor of the French city that hosted them while they worked on this project, the model is known as the Nice model.

In this model Uranus and Neptune accreted closer to the Sun, where they could more efficiently gain mass from the planetary nebula. Jupiter, Saturn, Uranus, and Neptune begin with near-circular orbits all between about 5.5 and 14 AU. Saturn is assumed to begin closer to Jupiter than their mutual 1:2 mean motion resonance (this resonance plays an important role in the model results). The small bodies in the Kuiper belt total 35 Earth masses to begin.

These small bodies would interact gravitationally with Uranus and Neptune in particular as the outermost giant planets. Every interaction that changes the orbit of the smaller body must also change the orbit of the larger body, by conservation of angular momentum. If the small body is thrown outward into a larger orbit, the big body moves a very small fraction of that distance into a smaller, faster orbit. Uranus and Neptune are too small to throw a majority of small bodies they encounter out of the solar system; most of the small bodies they encounter are thrown inward, toward Jupiter and Saturn and the Sun. Even bodies that Neptune throws outward generally do not leave the solar system but are eventually sent back inward to encounter Neptune again. The majority of bodies moving inward means that Neptune and Uranus move outward. They are forced to migrate into larger orbits, away from the Sun. Jupiter, on the other hand, has the giant gravity necessary to throw small bodies entirely outward, and so it migrates inward, closer to the Sun.

The team tested this model with many computer simulations, many with identical starting conditions. The process of planetary accretion and interaction of orbits, however, is stochastic: This means that the process contains a random component, depending upon probability. Exactly how the little bodies interact with the planets, at what rates and at what times, changes the progress and outcome of the simulation. Thus, every time the simulation is run the results are different.

As Neptune moved outward, it interacted with more and more Kuiper belt bodies, throwing many inward and others into stabilizing mean motion resonances. This progress continues slowly, over 350 million to 1.1 billion years, depending upon the simulation. At this point the slow progress of the movements ends and a brief period of highly energetic encounters begins, driven by the entrance of Jupiter and Saturn into their 1:2 mean motion resonance. This resonance makes the orbits of both these planets slightly more eccentric (elliptical). In turn, the orbits of Uranus and Neptune become far more eccentric, and Neptune's orbit drives into the Kuiper belt, scattering objects violently and at great rates.

A burst of Kuiper belt objects is thrown into the inner solar system and produces the Late Heavy Bombardment of the inner solar system, here a major spike in masses and rates of impact rather than the end of the accretionary tail, as other scientists have hypothesized. The Nice model predicts that 3.6×10^{18} lbs. (8×10^{18} kg) of material strikes the Moon during this Late Heavy Bombardment, a number consistent with observations of the lunar basins. In turn, the interactions among all these small bodies and the giant planets returns the orbits of the giant planets to lower eccentricity and ends the period of major interactions among the planets and the small bodies. Neptune settles at its current orbital distance of about 30 AU, having migrated outward by more than a factor of about two, from 14 AU.

The Kuiper belt is left highly depleted in numbers and in mass, and with many of the remaining bodies in stabilizing mean motion resonances. Uranus and Neptune are considerably farther from the Sun than they were during accretion.

The predictions of the computer simulations fit observations nicely, in terms of Uranus's and Neptune's orbits, the numbers and positions of Kuiper belt bodies, and the timing and size of the Late Heavy Bombardment. The simulations also predict that many small bodies would be scattered from the outer regions into the stable Lagrange points of Jupiter's orbit, where the Trojan asteroids now orbit along with the giant planet. In the simulations, 1 million objects with a total mass of 0.00003 times the mass of the Earth are trapped as Trojan asteroids in Jupiter's orbit. Observations show 160,000 asteroids with radius greater than 0.6 mile (1 km) with a total mass of 0.00001 times the mass of the Earth. The simulations are therefore in very good agreement with observations—in the world of complex computer modeling of stochastic systems, this is excellent agreement.

The agreement between the simulations and observations on such a range of topics has raised considerable interest in the scientific community and has led to many references to the articles written by the team, related studies by other groups, and a general assessment that this is the best model for evolution in our early solar system. The compelling ending conditions of the Nice model simulations, however, are dependent upon initial conditions in the simulations, even taking into consideration the stochastic nature of the problem.

The scientists behind the Nice model are continuing to work on the problem, focusing on where the giant planets begin in the simulation. A wide variety of initial orbits for the giant planets give the same satisfactory final simulation results for the giant planets and the Kuiper belt. Different starting orbits for the giant planets, however, lead to interesting events in the middle of the simulation runs: In some models, Saturn jumps over Uranus's orbit to lie between Uranus and Neptune, and then later in the same simulation, gravitational interaction with Jupiter makes Uranus jump back. Indeed, this may have happened; there is no observation today that can rule it out.

A further concern is the fates of the small terrestrial planets. When they are included in Nice model simulations, they are excited into highly eccentric and inclined orbits, inconsistent with what we see today, and also potentially unstable. A complete dynamical model for the solar system must include the terrestrial planets, or at least be consistent with what is now observed. In the latest simulations, the terrestrial planets are preserved in only 6 percent of the runs.

A next-generation Nice model is now being produced. In this model as in the original the four giant planets formed within 15 to 20 AU, a wide variety of arrangements within that rough constraint still result in an outcome consistent with observations. Jupiter can begin between 5 and 6 AU, Saturn between 8 and 9, Neptune between 11 and 12, and Uranus between 16 and 17. Beyond the giant planets, the disk of smaller material extends to 30 AU. Since most of the work happened in Boulder, Colorado, at the Southwest Research Institute, the new model may become known as the Boulder model.

PART THREE
BEYOND THE KUIPER BELT

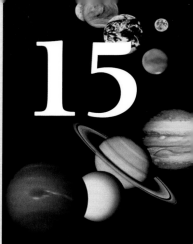

The Oort Cloud

By the 1900s humankind had good records on cometary appearances reaching back 200 to 300 years. Some comets were recognized as repeat visitors to the inner solar system and had earned names, and their appearances could be predicted. Scientists also noticed that some comets were new; that is, they were not recognized as having come through the inner solar system previously and so were assumed to have very long periods. Based somewhat arbitrarily on the length of time of recordkeeping, short-period comets are defined as those with periods less than 200 years, and long-period comets are those with periods over 200 years.

When Dutch astronomer Jan Oort could find no other explanation for long-period comets, he suggested in 1950 the existence of a huge swarm of comets surrounding the solar system. He had noted that comets approach the Earth from all angles, that is, their orbits are not confined to the ecliptic plane as the planets' orbits are. Oort also had enough data from watching comets pass the Earth to calculate their orbits, and he found from his calculations that comets with long periods (200 years or longer) had perihelia about 50,000 AU from the Sun. His hypothesized swarm of comets reaching out from the outer solar system was named the Oort cloud. At the distance that they orbit in the Oort cloud these bodies do not develop the distinctive cometary tail caused by the Sun, and they are thought to be reddened icy bodies similar to those in the Kuiper belt. The distance to the inner edge of the Oort cloud is disputed; some scientists believe that the innermost Oort cloud bodies reach as close as 70 AU from the Sun at their perihelia, while other scientists believe that the true Oort cloud begins vastly farther away, at about 1,000 AU, and any bodies closer to the Sun are a distinct population, perhaps a new outer Kuiper belt population, or perhaps a separate inner Oort cloud population. The comets' aphelia are thought to reach 30,000 to 60,000 AU from the Sun (some scientists say they reach to as much as 100,000 AU). Proxima Centauri is the star closest to the Sun, 4.2 light-years or about 265,000 AU away, so some Oort cloud objects travel a quarter to a third of the way to the nearest stars.

This set has progressed from the Sun outward, studying objects starting with Mercury and moving out through the planets and finally to the Kuiper belt. With the Oort cloud, this set has reached the extreme edges of the solar system, where interactions with other stars and in fact the rest of the galaxy becomes important. This solar system is a part of a spiral galaxy called the Milky Way, which contains perhaps 100 billion stars. Stars rotate around the center of the Milky Way much as the arms of a pinwheel rotate around its center. At 30,000 to 60,000 AU, the comet's aphelia are so far from the solar system that the Sun's gravity is weak and the comets' orbits are perturbed by passing stars, giant molecular clouds, and even the gravitational tides of the galactic disk and core. These gravitational perturbations sometimes divert the comets' orbits such that they pass into the inner solar system and are seen from Earth as long-term comets.

The plane in which the arms of the Milky Way rotate is called the galactic plane. In its movements through the galaxy, the solar system oscillates above and below the galactic plane, on a period of about 60 million years. Cometary flux from the Oort cloud is expected to vary by a factor of four over that period, as the galactic plane exerts varying force on the bodies in the Oort cloud. The maximum cometary flux is expected to occur just after the solar system passes through the galactic plane. The solar system has just passed through the galactic plane, so cometary flux should now be at a local maximum.

The shape of the Oort cloud is thought to be influenced by the solar system's travel through the galaxy and by the mass of the galaxy around us. The cloud is probably not truly round, but elongated toward the galactic center. The long radius of the Oort cloud is about 100,000 AU, and the radius of the short axis about 80,000 AU.

Random close approaches of stars or giant molecular clouds (the birthplaces of stars) can cause shorter-term, much more intense spikes in cometary passages than does passage through the galactic plane. Over the history of the solar system, an estimated 10,000 stars have passed within a few hundred thousand AU of the Sun at an average speed of 25 miles per second (40 km/sec), while the Sun is moving at only 10 to 12 miles per second (16 to 20 km/sec) relative to the galactic center. These passing stars have immense gravity fields and can pull distant Oort cloud bodies away from the gravity of the Sun, which in those circumstances actually can be much farther from the Oort cloud body than the passing alien star is, or perturb their orbits so that the bodies fall toward the Sun. Models show that about 10 percent of the Oort cloud may have been lost to passing stars in this fashion. Such a passing star could perturb a swarm on the order of 10^6 comets toward the Earth over a period of about 1 million years. Such a swarm has been suggested as a mechanism for extinctions on Earth. It is unlikely, however, that such an event would occur more frequently than once every 300 million to 500 million years, and so should have occurred only about nine to 15 times in Earth history.

When Oort cloud comets are perturbed and travel into the planetary system, the gravitational fields of the planets further perturb their orbits. Scientists calculate that roughly half the comets are ejected from the solar system into interplanetary space during their first pass into the planetary system. Most of the rest are captured into tighter orbits, and only about 5 percent are returned to the Oort cloud. Some smaller fraction falls into the Sun or collides with a planet: 10 to 25 percent of craters on the Earth and Moon larger than six miles (10 km) are thought to be created by long-period comets. About 10 new long-period comets are discovered each year, of which two or three pass inside the orbit of the Earth. The average successful long-period comet makes about five passes through the inner solar system, after which the majority are probably ejected from the solar system. Based on computer modeling, long-period comets have an average lifetime of about 600,000 years between their first and last passages through the inner solar system.

Because there should have been very little solar-system material at the distances from the Sun where the Oort cloud now lies, some scientists think that these long-term comets originated as icy planetesimals in the vicinity of the gas giant planets. The gravitational fields of the growing gas giants expelled the smaller planetesimals beyond Pluto and into the Oort cloud. Other icy planetesimals that originated in the vicinity of Pluto were far enough from the gas giants to remain where they formed, and they now make up the Kuiper belt.

The formation of the Oort cloud is not well understood, and neither in fact is a comet's evolution. Several prominent comets, including 1P Halley, 109P Swift-Tuttle, and 55P Tempel-Tuttle, have visits to the inner solar system that can be traced back 2,000 years through the records of different civilizations. The brightness of these comets over 20 or more trips around the Sun is something of a mystery: Should the cometary bodies have lost much of their ices and gases by now? The details of the process of creating cometary tails are becoming better understood because of recent space missions to comets, and perhaps their longevity will be explained.

Kuiper belt bodies may be the unprocessed material from the initial solar nebula, but the far more distant Oort cloud bodies may actually be more processed than their cousins in the Kuiper belt. Oort cloud bodies may be affected by processes caused by galactic cosmic rays: These high-energy rays irradiate the bodies, can energize particles on their surface sufficiently that they sputter away into space, and may cause chemical reactions in the Oort cloud body material. Oort cloud bodies may also be heated by passing stars and nearby supernovae more intensely than they are by the Sun. Their surfaces may also be coated with interstellar gas and dust with compositions different from the dust encountered within the Sun's magnetosphere. These processes may form a permanent nonvolatile crust on the bodies, through which gas jets erupt if the body travels into the inner solar system and becomes a comet.

Some scientists estimate that 10^{12} to 10^{13} bodies exist in the Oort cloud. Though the population of the Oort cloud was estimated using mathematical modeling techniques, its mass is much less certain. Even the mass of comet Halley is not known with any precision. Its mass has to be calculated based on its size and its density, and density estimates range from 200 to 1,200 kg/m^3. The range of masses of all comets is even less well known. Masses have been estimated by using the distribution of cometary magnitudes (brightnesses), and also from estimates of masses from the Kuiper belt. A plausible average density is about 600 kg/m^3, and an average mass for a comet is about 10^{13} kg. Using these estimates, the Oort cloud contains about 20 to 40 Earth masses.

In 2004 the press carried a breathless story about the discovery of a new planet in the solar system. The planetary science community, however, was far less shocked: This was simply a large, distant object (the press announcement occurred the week of the annual Lunar and Planetary Science Conference, the largest planetary science conference in the world, but almost no mention of the discovery was made at the conference). The new object is named Sedna after the Inuit goddess of the ocean (it was originally designated 2003 VB_{12}). Sedna was discovered by the California Institute of Technology professor Michael Brown in November 2003 during a two-and-a-half-year sky survey for Kuiper belt objects. Each night the scientists use telescopes at the Palomar Observatory in Southern California to take three photos of a small region of the sky, one hour apart. They look for objects that move between the images, and then check to see if those objects are already known. The patches of sky being examined are exceptionally tiny; Brown describes them as appearing like the size of a pinhead held at arm's length. Brown and his team had examined 15 percent of the sky in this manner before finding Sedna.

Soon after he found it, Brown realized that it was far more distant than the Kuiper belt, which reaches from 30 to 49 AU, where it has an abrupt outer edge. Sedna is currently at about 90 AU distance from the Sun and will reach its estimated perihelion of 76 AU in 2075. Sedna is thus thought to be the first identified object in the Oort cloud, which until 2005 was truly a theoretical construct to explain the orbits of comets. Sedna is more than twice as far from the Sun as the next closest object yet identified; the next farthest object now known is the Kuiper belt object 1999 CL_{119}, which has a perihelion distance from the Sun of 46.6 AU. Comparisons of the orbits of the inner and outer planets, the Kuiper belt, Sedna, and the Oort cloud are shown in this image. Though Sedna is not a planet, it is actually more exciting as the first known object orbiting within the Oort cloud. Jan Oort himself, sadly, died in 1992 and so did not survive to witness the discovery that validated his excellent theories.

Sedna is nearing its perihelion now, but its orbit has an eccentricity of 0.84 and carries the object to an aphelion of more than 800 AU from the Sun! This remarkable distance is 10 times farther than the Sun's magnetic field and solar wind reach. The distance carries Sedna into interstellar space, out of the field of influence of everything in the solar system except the Sun's gravity. Sedna orbits the Sun once every 10,500 years in an orbit with 12 degrees inclination from the ecliptic.

The Sedna discovery image: In three images taken on November 14, 2003, Sedna was identified by the slight shift in position visible between the pictures. (NASA/Michael Brown/CalTech)

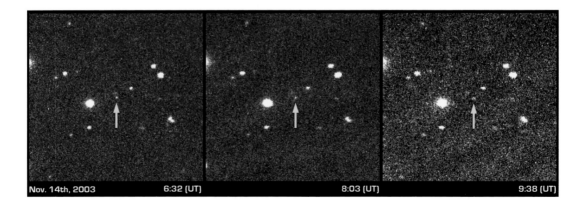

Nov. 14th, 2003   6:32 (UT)   8:03 (UT)   9:38 (UT)

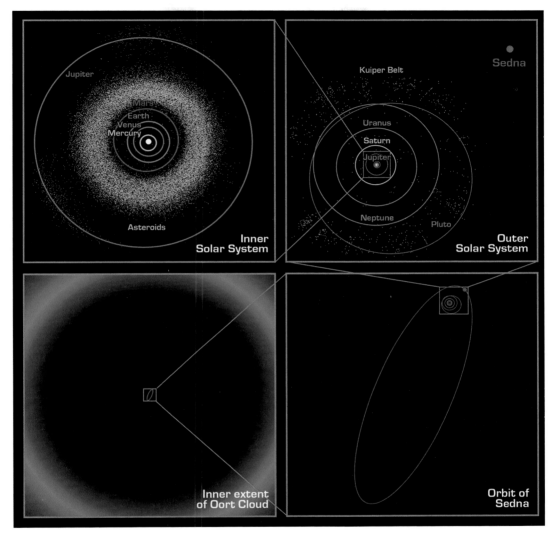

Moving clockwise from the upper left, each successive panel in this image zooms out from the inner solar system to the inner Oort cloud, showing Sedna's orbit outside the Kuiper belt but well inside the canonical Oort cloud. (NASA/CalTech)

Sedna's radius is about 500 miles (800 km), and so it is smaller than Pluto (radius of 715 miles or 1,151 km) and the Moon (radius of 1,079 miles, or 1,737 km). Its color is deep red, for unknown reasons, though its brightness is attributed to methane ice. It never warms above −400°F (−240°C) and so may represent some of the most primordial and unprocessed material in the solar system. It is so distant that despite its brightness it cannot be seen through any amateur telescope.

Sedna's brightness fluctuates on a 20-day cycle. Brown and his colleagues believe that the fluctuation is caused by bright and dark parts of Sedna's surface moving past as the object rotates on a 20-day cycle. A 20-day rotation is exceptionally slow: Almost every other body in the solar system rotates in hours or a day or so, except those bodies that are being influenced by the gravity of a moon (for example, Pluto rotates once every six days) or by the Sun if it is very close (for example, Mercury). Sedna's slow rotation was thought to be caused by a moon, but none has been seen.

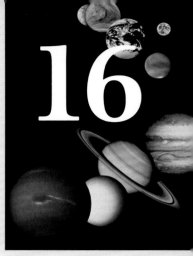

Conclusions:
The Known and
the Unknown

Voyager 2 is the only space mission to have flown close to either Uranus or Neptune. It flew by Uranus in January 1986, after nine years in space. *Voyager 2* took high-resolution images of Uranus's atmosphere, the ring system, and 16 moons, 11 of which were discovered by the *Voyager 2* mission. The mission flew by Neptune in August 1989, after 12 years in space. Calculations on where Neptune would be were slightly incorrect, and to the surprise of NASA scientists, *Voyager 2* arrived four minutes and 45 seconds earlier than expected. *Voyager 2* discovered Neptune's moons Proteus, Larissa, Despina, Galatea, Thalassa, and Naiad. The mission also confirmed the existence of rings around Neptune, and took images detailed enough to allow the rings to be named and described. In fact, almost everything known about Uranus and Neptune came from the Voyager 2 mission or later interpretations or embellishments of its data. Only recently, with new ground-based observational techniques and with the use of the *Hubble Space Telescope,* have improved data for Uranus and Neptune been acquired.

Pluto is the only one of the original planets that has not been visited by a spacecraft. In spite of its extreme remoteness and the complete lack of data from space missions, Pluto is the subject of intense study. Its position in the Kuiper belt makes it an unusually interesting object in that it can give greater insight into solar system dynamics and evolution than can closer planets. A spacecraft called *New Horizons,* launched on January 19, 2006, is scheduled to reach Pluto and Charon in July 2015. The spacecraft would then head deeper into the Kuiper belt to study one or more of the bodies in that vast region, at least a billion miles beyond Neptune's orbit. The mission includes instrumentation for visible-wavelength surface mapping, infrared and ultraviolet imaging to study surface composition and atmosphere, radiometry, and solar wind measurements.

Interest in the outermost solar system, from the most distant Kuiper belt objects to the Oort cloud, has reached new heights with the discovery of Sedna. The understanding scientists have reached about this region of the solar system lies almost entirely in the realm of theory, with so few measurements as to be almost negligible. In terms of volume, this part of the solar system is far larger than the better-known portions of the solar system, and research into these intriguing and mysterious reaches is ongoing despite the intense challenges of distance and the small sizes of the objects being sought. Some of the most pressing questions about the outer solar system are the following:

1. **How does the Uranian atmosphere smooth out the seasonal extremes of solar heating?**

   Uranus's rotation axis lies almost flat in its orbital plane and remains pointed in a fixed direction as the planet orbits around the Sun. As a result, for one-quarter of its orbit the Sun shines more or less directly on its north pole (and the south pole remains in complete darkness), and then for the next quarter, the Sun shines on the planet's rotating equator (which then experiences day and night), and for the next quarter, almost exclusively on the planet's south pole. Each of these quarter orbits lasts about 20 years. The long periods of heating one pole and freezing the other in complete darkness obviously produce far more extreme seasons than are experienced on Earth.

   Extreme heating on one side of the planet with cooling on the other should produce violent storms as heat is transferred from the hot pole to the cold, and temperature gradients on the planet should be easily measurable. Contrary to these simple models, surface temperatures measured on Uranus are relatively small. The *Hubble Space Telescope* has spotted giant storms on Uranus. Storms are an effect of heat transport, but Uranus apparently has some physical method beyond storms for equalizing heat across the planet with supreme efficiency. Despite the fact that the planet's most extreme heating occurs on its poles and so one would expect flow patterns from pole to pole, the planet displays bands of movement parallel to its equator, just as Jupiter and Saturn do. Material flowing south from the north pole or north from the south pole is thought to be deflected by the *Coriolis force* of the spinning planet, which acts to curve the flow into bands parallel to the equator. This simple physical explanation does not help explain the efficient heat flow from pole to pole. The Uranian atmosphere obviously has fascinating hidden processes. Of course, observers have only been able to take good measurements of the temperature of Uranus over the last few decades. Could understanding be limited by obtaining measurements only during one Uranian season? Could better-resolved images and better-resolved temperature measurements provide data that would help explain the efficient heat flow, or will better understanding have to await missions that reach Uranus during later seasons in its long orbit?

2. **Why is Uranus's internal heat source so small, or the planet so well insulated, when compared to other gas giant planets?**
   Uranus's heat flux through its surface is another anomaly that sets it apart from other planets: The other planets leak heat through their surfaces, releasing energy from internal radioactive decay or other internal processes. Heat loss through Uranus's surface is undetectable. Unlike Jupiter and Saturn, Uranus does not give off more heat than it receives from the Sun. Planets produce heat through decay of radioactive elements, and in the cases of Jupiter and Saturn, through the potential energy of sinking helium rain

in their interiors. Helium condenses from a gas and rains into the deeper interior of the planets, releasing energy as it sinks. Uranus, however, does not seem to be losing heat at all. Perhaps its relative lack of helium removes the main gas planet mechanism for producing internal heat. Measuring heat loss from the interior of a planet is the best, and often the only, way to calculate its internal temperatures. Because Uranus is not losing any heat, its internal temperatures are completely unknown.

Neptune, on the other hand, has measurable heat flux from its surface and temperature differences over its surface that are more similar to the other gas giant planets. Weather patterns are more visible on its surface than on the largely featureless Uranus. Neptune's characteristics are consistent with a planet losing heat from its interior, which is also consistent with the planet's robust magnetic field. Uranus's lack of heat flow is also a conundrum for the existence of its magnetic field, since fields appear to require heat loss from the planet.

3. **What is the source of energy that heats both Neptune's and Uranus's upper atmospheres to such high temperatures?**

In Uranus's stratosphere the temperature rises from the low of about −364°F (−220°C) at the tropopause to a high of perhaps 890°F (477°C) in the exosphere (the uppermost layer of its atmosphere). Neptune's exosphere appears to be slightly cooler, reaching only about 620°F (327°C). The extreme warmth of the upper atmospheres is not understood; there is no explanation from solar heating or from energy from the planet's interior.

Because Uranus's outer atmosphere is so calm, what methane there was in the outer atmosphere freezes and settles back into the lower atmosphere with relatively little remixing. There are few hydrocarbons for solar radiation to turn into smog, and the outer atmosphere remains clear. The atmosphere cannot retain heat through greenhouse processes when it is so clear and free of hydrocarbons. Neptune's outer atmosphere has a higher haze content than does

Uranus's, but Neptune's outer atmosphere is slightly cooler than Uranus's. Ultraviolet radiation from the Sun cannot heat the outer atmosphere to these temperatures. The other gas giant planets have similarly hot exospheres, and on Jupiter the heat is thought to come from ionizing reactions tied to Jupiter's magnetic field and its auroras. Perhaps a similar process is at work on Uranus and Neptune.

4. **Why are radio emissions from Uranus and Neptune oriented and why do they occur in bursts, when the Earth's and Saturn's radio emissions are controlled by the solar wind and are therefore smooth and not attached to the planet's rotation?**
   Uranus and Neptune emit radio waves with frequencies in the 100 to 1,000 kHz range, both in bursts and continuously. These emissions are not similar to those from the other gas giant planets. The emissions are oriented and rotate with the planet; at present, the strongest emissions from Uranus are aimed away from the Earth. The emissions seem to originate in the upper atmosphere along magnetic and auroral field lines and are thought to originate from electrons spiraling along field lines. Interactions among the extreme tilt of Uranus's rotation axis, ions produced by its moons, and the solar wind may account for the intermittent bursts of radio emissions. Neptune's magnetic field is highly tilted, though its rotation axis is not, so perhaps these offsets have a similar effect to the tilt of Uranus's axis.

5. **Why do Uranus and Neptune have so much less helium and hydrogen than Jupiter and Saturn? Light elements are thought to have been more enriched in the outer portions of the solar nebula.**
   The outer portions of Uranus and Neptune consist of gaseous helium and hydrogen, as do their other neighbors in the outer solar system, but they also have methane, ammonia, and water in their atmospheres. The bulk of each planet is made up of a range of ices, and the gaseous helium and hydrogen together make up only about 15 percent of

the planets, far less than Jupiter and Saturn are thought to have. Was the excess helium and hydrogen removed from the early planets in some way? Could the planets have formed in a region of lower helium and migrated to their current locations (this is unlikely, due to the huge masses of Jupiter and Saturn, which would disrupt other migrating planets)? Was the solar nebular much less homogeneous than has been previously assumed?

6. **How many large bodies are there orbiting near and beyond Pluto, and how large might they be?**
   Objects as large as Mars may exist in the outer Kuiper belt. Continuing surveys and new surveys with new instruments will search the outer belt in the hopes of finding more large bodies, which can further add to the understanding of the composition and distribution of the solar nebula.

7. **What orbits in the region between the Kuiper belt and the Oort cloud?**
   The seeming edge of the Kuiper belt at 49 AU is a source of interest because there is no clear reason for its existence. The Taiwan American Occultation Survey (TAOS) is a joint effort of scientists at the Acadamia Sinica, Institute of Astronomy and Astrophysics in Taiwan, NASA Ames, the National Central University in Taiwan, the Lawrence Livermore National Laboratory, the University of California at Berkeley, the University of Pennsylvania, and Yonsei University in Korea. TAOS has three large telescopes installed in the mountains of Taiwan, and by watching for star occultations, scientists there hope to catalog Kuiper belt objects as small as one or two miles (2 or 3 km) in diameter.
   By mapping these smaller objects, the teams hope to clarify the radial extent of the Kuiper belt, and demonstrate whether or not there is a clear edge at 49 AU. Understanding the distribution of sizes in the Kuiper belt may also help modeling efforts to explain its shape and extent. The proposed Large Synoptic Survey Telescope, a joint effort by many partners in the United States to build a 28-foot (8.4-meter) telescope to make continuous deep-field sur-

veys, would also greatly enhance knowledge of the outer solar system.

8. **What are Kuiper belt and Oort cloud bodies made of?**

The range of colors and albedos in the Kuiper belt is unexpected. This range may be caused by a continuum between aged, irradiated surfaces and surfaces variably disturbed by collisions, with fresher, deeper material showing, this collision theory cannot explain all the data. Pluto and Charon, though they orbit each other closely and may thus have originated at similar radial distances from the Sun, are proven to have different compositions. Pluto is rich in nitrogen, methane, and carbon monoxide, while Charon has none of those but appears to contain significant ammonia. Both bodies have a rocky component and a large fraction of water ice. A wide variation in compositions is unexpected for the Kuiper belt and remains to be explained by theories of solar system evolution. The one possible Oort cloud body seen to date is intensely reddened, indicating a possible enrichment in organic molecules. Further information on these most distant bodies may entirely change the current view that those bodies must consist almost entirely of ices.

9. **Can the Kuiper belt and Oort cloud be used to understand circumstellar dust disks?**

Dust rings around other stars have been observed from Earth telescopes, but their formation and maintenance is not understood. The two Voyager missions detected dust in the Kuiper belt as they passed through. This dust may be original to the solar nebula, and it may also be produced by collisions among the Kuiper belt objects. The tiniest dust grains, smaller than one micron, are blown out of the solar system by the solar wind. The rest slowly spiral toward the Sun as they lose angular momentum in collisions with other dust grains, and through the pressures of absorbing and re-emitting solar radiation. Large grains (more than 50 microns or so) probably break up in further collisions. About one-fifth of the dust grains are expected to reach the

Sun, and the rest are thought to be expelled from the solar system. Are there collections of bodies like the Kuiper belt around these distant stars, or do stellar dust clouds mark a transient, earlier state in solar system evolution?

All of the vast outer solar system is little known and a subject of great curiosity. Many unanswered questions await future space missions and the new data they can collect. The Oort cloud in particular is a kind of final frontier in solar system science because of how little is known about it and because of its extreme distances from the Sun. Oort cloud bodies still orbit the Sun and are controlled to a large degree by its gravity field, but they also experience the interstellar medium and its different radiation environment, and are perturbed by passing stars and molecular clouds, almost unimaginably exotic to humankind in the warm inner solar system. The tantalizing glimpses obtained through the brief visits of long-term comets into the inner solar system serve to motivate further research, through which scientists hope to better understand the processes that formed the solar system in its first few million years of existence.

APPENDIX 1:

Units and Measurements

FUNDAMENTAL UNITS

The system of measurements most commonly used in science is called both the SI (for Système International d'Unités) and the International System of Units (it is also sometimes called the MKS system). The SI system is based upon the metric units meter (abbreviated m), kilogram (kg), second (sec), kelvin (K), mole (mol), candela (cd), and ampere (A), used to measure length, time, mass, temperature, amount of a substance, light intensity, and electric current, respectively. This system was agreed upon in 1974 at an international general conference. There is another metric system, CGS, which stands for centimeter, gram, second; that system simply uses the hundredth of a meter (the centimeter) and the hundredth of the kilogram (the gram). The CGS system, formally introduced by the British Association for the Advancement of Science in 1874, is particularly useful to scientists making measurements of small quantities in laboratories, but it is less useful for space science. In this set, the SI system is used with the exception that temperatures will be presented in Celsius (C), instead of Kelvin. (The conversions between Celsius, Kelvin, and Fahrenheit temperatures are given below.) Often the standard unit of measure in the SI system, the meter, is too small when talking about the great distances in the solar system; kilometers (thousands of meters) or AU (astronomical units, defined below) will often be used instead of meters.

How is a unit defined? At one time a "meter" was defined as the length of a special metal ruler kept under strict conditions of temperature and humidity. That perfect meter could not be measured, however, without changing its temperature by opening the box, which would change its length, through thermal expansion or contraction. Today a meter is no longer defined according to a physical object; the only fundamental measurement that still is defined by a physical object is the kilogram. All of these units have had long and complex histories of attempts to define them. Some of the modern definitions, along with the use and abbreviation of each, are listed in the table here.

FUNDAMENTAL UNITS			
Measurement	**Unit**	**Symbol**	**Definition**
length	meter	m	The meter is the distance traveled by light in a vacuum during 1/299,792,458 of a second.
time	second	sec	The second is defined as the period of time in which the oscillations of cesium atoms, under specified conditions, complete exactly 9,192,631,770 cycles. The length of a second was thought to be a constant before Einstein developed theories in physics that show that the closer to the speed of light an object is traveling, the slower time is for that object. For the velocities on Earth, time is quite accurately still considered a constant.
mass	kilogram	kg	The International Bureau of Weights and Measures keeps the world's standard kilogram in Paris, and that object is the definition of the kilogram.
temperature	kelvin	K	A degree in Kelvin (and Celsius) is 1/273.16 of the thermody- namic temperature of the triple point of water (the temper- ature at which, under one atmosphere pressure, water coexists as water vapor, liquid, and solid ice). In 1967, the General Conference on Weights and Measures defined this temperature as 273.16 kelvin.

Measurement	Unit	Symbol	Definition
amount of a substance	mole	mol	The mole is the amount of a substance that contains as many units as there are atoms in 0.012 kilogram of carbon 12 (that is, Avogadro's number, or 6.02205×10^{23}). The units may be atoms, molecules, ions, or other particles.
electric current	ampere	A	The ampere is that constant current which, if maintained in two straight parallel conductors of infinite length, of negligible circular cross section, and placed one meter apart in a vacuum, would produce between these conductors a force equal to 2×10^{-7} newtons per meter of length.
light intensity	candela	cd	The candela is the luminous intensity of a source that emits monochromatic radiation with a wavelength of 555.17 nm and that has a radiant intensity of 1/683 watt per steradian. Normal human eyes are more sensitive to the yellow-green light of this wavelength than to any other.

Mass and weight are often confused. Weight is proportional to the force of gravity: Your weight on Earth is about six times your weight on the Moon because Earth's gravity is about six times that of the Moon's. Mass, on the other hand, is a quantity of matter, measured independently of gravity. In fact, weight has different units from mass: Weight is actually measured as a force (newtons, in SI, or pounds, in the English system).

The table "Fundamental Units" lists the fundamental units of the SI system. These are units that need to be defined in order to make other measurements. For example, the meter and the second are fundamental units (they are not based on any other units). To measure velocity, use a derived unit, meters per second (m/sec), a combination of fundamental units. Later in this section there is a list of common derived units.

The systems of temperature are capitalized (Fahrenheit, Celsius, and Kelvin), but the units are not (degree and kelvin). Unit abbreviations are capitalized only when they are named after a person, such as K for Lord Kelvin, or A for André-Marie Ampère. The units themselves are always lowercase, even when named for a person: one newton, or one N. Throughout these tables a small dot indicates multiplication, as in N m, which means a newton (N) times a meter (m). A space between the symbols can also be used to indicate multiplication, as in N · m. When a small letter is placed in front of a symbol, it is a prefix meaning some multiplication factor. For example, J stands for the unit of energy called a joule, and a mJ indicates a millijoule, or 10^{-3} joules. The table of prefixes is given at the end of this section.

COMPARISONS AMONG KELVIN, CELSIUS, AND FAHRENHEIT

One kelvin represents the same temperature difference as 1°C, and the temperature in kelvins is always equal to 273.15 plus the temperature in degrees Celsius. The Celsius scale was designed around the behavior of water. The freezing point of water (at one atmosphere of pressure) was originally defined to be 0°C, while the boiling point is 100°C. The kelvin equals exactly 1.8°F.

To convert temperatures in the Fahrenheit scale to the Celsius scale, use the following equation, where F is degrees Fahrenheit, and C is degrees Celsius:

$$C = (F - 32)/1.8.$$

And to convert Celsius to Fahrenheit, use this equation:

$$F = 1.8C + 32.$$

To convert temperatures in the Celsius scale to the Kelvin scale, add 273.16. By convention, the degree symbol (°) is used for Celsius and Fahrenheit temperatures but not for temperatures given in Kelvin, for example, 0°C equals 273K.

What exactly is temperature? Qualitatively, it is a measurement of how hot something feels, and this definition is so easy to relate to that people seldom take it further. What is really happening in a substance as it gets hot or cold, and how does that change make temperature? When a fixed amount of energy is put into a substance, it heats up by an amount depending on what it is. The temperature of an object, then, has something to do with how the material responds to energy, and that response is called entropy. The entropy of a material (entropy is usually denoted S) is a measure of atomic wiggling and disorder of the atoms in the material. Formally, temperature is defined as

$$\frac{1}{T} = \left(\frac{dS}{dU} \right)_{N,}$$

meaning one over temperature (the reciprocal of temperature) is defined as the change in entropy $(dS,$ in differential notation) per change in energy $(dU),$ for a given number of atoms (N). What this means in less technical terms is that temperature is a measure of how much heat it takes to increase the entropy (atomic wiggling and disorder) of a substance. Some materials get hotter with less energy, and others require more to reach the same temperature.

The theoretical lower limit of temperature is $-459.67°F$ $(-273.15°C,$ or 0K), known also as absolute zero. This is the temperature at which all atomic movement stops. The Prussian physicist Walther Nernst showed that it is impossible to actually reach absolute zero, though with laboratory methods using nuclear magnetization it is possible to reach 10^{-6}K $(0.000001K)$.

USEFUL MEASURES OF DISTANCE

A *kilometer* is a thousand meters (see the table "International System Prefixes"), and a *light-year* is the distance light travels in a vacuum during one year (exactly 299,792,458 m/sec, but commonly rounded to 300,000,000 m/sec). A light-year, therefore, is the distance that light can travel in one year, or:

$$299{,}792{,}458 \text{ m/sec} \times 60 \text{ sec/min} \times 60 \text{ min/hr} \times$$
$$24 \text{ hr/day} \times 365 \text{ days/yr} = 9.4543 \times 10^{15} \text{ m/yr}.$$

For shorter distances, some astronomers use light minutes and even light seconds. A light minute is 17,998,775 km, and a light second is 299,812.59 km. The nearest star to Earth, Proxima Centauri, is 4.2 light-years away from the Sun. The next, Rigil Centaurs, is 4.3 light-years away.

An *angstrom* (10^{-10}m) is a unit of length most commonly used in nuclear or particle physics. Its symbol is Å. The diameter of an atom is about one angstrom (though each element and isotope is slightly different).

An astronomical unit (AU) is a unit of distance used by astronomers to measure distances in the solar system. One astronomical unit equals the average distance from the center of the Earth to the center of the Sun. The currently accepted value, made standard in 1996, is 149,597,870,691 meters, plus or minus 30 meters.

One kilometer equals 0.62 miles, and one mile equals 1.61 kilometers.

The table on page 197 gives the most commonly used of the units derived from the fundamental units above (there are many more derived units not listed here because they have been developed for specific situations and are little-used elsewhere; for example, in the metric world, the curvature of a railroad track is measured with a unit called "degree of curvature," defined as the angle between two points in a curving track that are separated by a chord of 20 meters).

Though the units are given in alphabetical order for ease of reference, many can fit into one of several broad categories: dimensional units (angle, area, volume), material properties (density, viscosity, thermal expansivity), properties of motion (velocity, acceleration, angular velocity), electrical properties (frequency, electric charge, electric potential, resistance, inductance, electric field strength), magnetic properties (magnetic field strength, magnetic flux, magnetic flux density), and properties of radioactivity (amount of radioactivity and effect of radioactivity).

(continues on page 200)

DERIVED UNITS		
Measurement	**Unit symbol (derivation)**	**Comments**
acceleration	unnamed (m/sec^2)	
angle	radian rad (m/m)	One radian is the angle centered in a circle that includes an arc of length equal to the radius. Since the circumference equals two pi times the radius, one radian equals 1/(2 pi) of the circle, or approximately 57.296°.
	steradian sr (m^2/ m^2)	The steradian is a unit of solid angle. There are four pi steradians in a sphere. Thus one steradian equals about 0.079577 sphere, or about 3282.806 square degrees.
angular velocity	unnamed (rad/sec)	
area	unnamed (m^2)	
density	unnamed (kg/m^3)	Density is mass per volume. Lead is dense, styrofoam is not. Water has a density of one gram per cubic centimeter or 1,000 kilograms per cubic meter.
electric charge or electric flux	coulomb C (A·sec)	One coulomb is the amount of charge accumulated in one second by a current of one ampere. One coulomb is also the amount of charge on 6.241506×10^{18} electrons.
electric field strength	unnamed [(kg·m)/ (sec^3·A) × V/m]	Electric field strength is a measure of the intensity of an electric field at a particular location. A field strength of one V/m represents a potential difference of one volt between points separated by one meter.
electric potential, or electromotive force (often called voltage)	volt V [(kg·m^2)/ (sec^3·A) = J/C = W/A]	Voltage is an expression of the potential difference in charge between two points in an electrical field. Electric potential is defined as the amount of potential energy present per unit of charge. One volt is a potential of one joule per coulomb of charge. The greater the voltage, the greater the flow of electrical current.

(continues)

DERIVED UNITS *(continued)*		
Measurement	**Unit symbol (derivation)**	**Comments**
energy, work, or heat	joule J [N·m (=kg·m²/sec²)]	
	electron volt eV	The electron volt, being so much smaller than the joule (one eV = 1.6×10^{-17} J), is useful for describing small systems.
force	newton N (kg·m/sec²)	This unit is the equivalent to the pound in the English system, since the pound is a measure of force and not mass.
frequency	hertz Hz (cycles/sec)	Frequency is related to wavelength as follows: kilohertz × wavelength in meters = 300,000.
inductance	henry H (Wb/A)	Inductance is the amount of magnetic flux a material pro- duces for a given current of electricity. Metal wire with an electric current passing through it creates a magnetic field; different types of metal make magnetic fields with different strengths and therefore have different inductances.
magnetic field strength	unnamed (A/m)	Magnetic field strength is the force that a magnetic field exerts on a theoretical unit magnetic pole.
magnetic flux	weber Wb [(kg·m²)/ (sec²·A) = V·sec]	The magnetic flux across a perpendicular surface is the product of the magnetic flux density, in teslas, and the surface area, in square meters.
magnetic flux density	tesla T [kg/(sec²·A) = Wb/m²]	A magnetic field of one tesla is strong: The strongest artificial fields made in laboratories are about 20 teslas, and the Earth's magnetic flux density, at its surface, is about 50 microteslas (μT). Planetary magnetic fields are sometimes measured in gammas, which are nanoteslas (10^{-9} teslas).
momentum, or impulse	unnamed [N·sec (= kg m/sec)]	Momentum is a measure of moving mass: how much mass and how fast it is moving.

Measurement	Unit symbol (derivation)	Comments
power	watt W [J/s (= $(kg\ m^2)/sec^3$)]	Power is the rate at which energy is spent. Power can be mechanical (as in horsepower) or electrical (a watt is produced by a current of one ampere flowing through an electric potential of one volt).
pressure, or stress	pascal Pa (N/m^2)	The high pressures inside planets are often measured in gigapascals (10^9 pascals), abbreviated GPa. ~10,000 atm = one GPa.
	atmosphere atm	The atmosphere is a handy unit because one atmosphere is approximately the pressure felt from the air at sea level on Earth; one standard atm = 101,325 Pa; one metric atm = 98,066 Pa; one atm ~ one bar.
radiation per unit mass receiving it	gray (J/kg)	The amount of radiation energy absorbed per kilogram of mass. One gray = 100 rads, an older unit.
radiation (effect of)	sievert Sv	This unit is meant to make comparable the biological effects of different doses and types of radiation. It is the energy of radiation received per kilogram, in grays, multiplied by a factor that takes into consideration the damage done by the particular type of radiation.
radioactivity (amount)	becquerel Bq	One atomic decay per second
	curie Ci	The curie is the older unit of measure but is still frequently seen. One Ci = 3.7×10^{10} Bq.
resistance	ohm Ω (V/A)	Resistance is a material's unwillingness to pass electric current. Materials with high resistance become hot rather than allowing the current to pass and can make excellent heaters.
thermal expansivity	unnamed (/°)	This unit is per degree, measuring the change in volume of a substance with the rise in temperature.

(continues)

	DERIVED UNITS *(continued)*	
Measurement	**Unit symbol (derivation)**	**Comments**
vacuum	torr	Vacuum is atmospheric pressure below one atm (one torr = 1/760 atm). Given a pool of mercury with a glass tube standing in it, one torr of pressure on the pool will press the mercury one millimeter up into the tube, where one standard atmosphere will push up 760 millimeters of mercury.
velocity	unnamed (m/sec)	
viscosity	unnamed [Pa·sec (= kg/ (m·sec)]	Viscosity is a measure of resistance to flow. If a force of one newton is needed to move one square meter of the liquid or gas relative to a second layer one meter away at a speed of one meter per second, then its viscosity is one Pa·s, often simply written Pas or Pas. The cgs unit for viscosity is the poise, equal to 0.1Pa·s.
volume	cubic meter (m³)	

(continued from page 196)

DEFINITIONS FOR ELECTRICITY AND MAGNETISM

When two objects in each other's vicinity have different electrical charges, an *electric field* exists between them. An electric field also forms around any single object that is electrically charged with respect to its environment. An object is negatively charged (-) if it has an excess of electrons relative to its surroundings. An object is positively charged (+) if it is deficient in electrons with respect to its surroundings.

An electric field has an effect on other charged objects in the vicinity. The field strength at a particular distance from an object is directly proportional to the electric charge of that object, in coulombs. The field strength is inversely proportional to the distance from a charged object.

Flux is the rate (per unit of time) in which something flowing crosses a surface perpendicular to the direction of flow.

An alternative expression for the intensity of an electric field is *electric flux density*. This refers to the number of lines of electric flux passing at right angles through a given surface area, usually one meter squared (1 m²). Electric flux density, like electric field strength, is directly proportional to the

INTERNATIONAL SYSTEM PREFIXES		
SI prefix	**Symbol**	**Multiplying factor**
exa-	E	10^{18} = 1,000,000,000,000,000,000
peta-	P	10^{15} = 1,000,000,000,000,000
tera-	T	10^{12} = 1,000,000,000,000
giga-	G	10^{9} = 1,000,000,000
mega-	M	10^{6} = 1,000,000
kilo-	k	10^{3} = 1,000
hecto-	h	10^{2} = 100
deca-	da	10 = 10
deci-	d	10^{-1} = 0.1
centi-	c	10^{-2} = 0.01
milli-	m	10^{-3} = 0.001
micro-	μ or u	10^{-6} = 0.000,001
nano-	n	10^{-9} = 0.000,000,001
pico-	p	10^{-12} = 0.000,000,000,001
femto-	f	10^{-15} = 0.000,000,000,000,001
atto-	a	10^{-18} = 0.000,000,000,000,000,001

A note on nonmetric prefixes: In the United States, the word billion means the number 1,000,000,000, or 10^{9}. In most countries of Europe and Latin America, this number is called "one milliard" or "one thousand million," and "billion" means the number 1,000,000,000,000, or 10^{12}, which is what Americans call a "trillion." In this set, a billion is 10^{9}.

NAMES FOR LARGE NUMBERS

Number	American	European	SI prefix
10^9	billion	milliard	giga-
10^{12}	trillion	billion	tera-
10^{15}	quadrillion	billiard	peta-
10^{18}	quintillion	trillion	exa-
10^{21}	sextillion	trilliard	zetta-
10^{24}	septillion	quadrillion	yotta-
10^{27}	octillion	quadrilliard	
10^{30}	nonillion	quintillion	
10^{33}	decillion	quintilliard	
10^{36}	undecillion	sextillion	
10^{39}	duodecillion	sextilliard	
10^{42}	tredecillion	septillion	
10^{45}	quattuordecillion	septilliard	

This naming system is designed to expand indefinitely by factors of powers of three. Then, there is also the googol, the number 10^{100} (one followed by 100 zeroes). The googol was invented for fun by the eight-year-old nephew of the American mathematician Edward Kasner. The googolplex is 10^{googol}, or one followed by a googol of zeroes. Both it and the googol are numbers larger than the total number of atoms in the universe, thought to be about 10^{80}.

charge on the object. But flux density diminishes with distance according to the inverse-square law because it is specified in terms of a surface area (per meter squared) rather than a linear displacement (per meter).

A *magnetic field* is generated when electric charge carriers such as electrons move through space or within an electrical conductor. The geometric shapes of the magnetic flux lines produced by moving charge carriers (electric current) are similar to the shapes of the flux lines in an electrostatic field. But there are differences in the ways electrostatic and magnetic fields interact with the environment.

Electrostatic flux is impeded or blocked by metallic objects. *Magnetic flux* passes through most metals with little or no effect, with certain exceptions, notably iron and nickel. These two metals, and alloys and mixtures containing them, are known as ferromagnetic materials because they concentrate magnetic lines of flux.

Magnetic flux density and *magnetic force* are related to *magnetic field strength*. In general, the magnetic field strength diminishes with increasing distance from the axis of a magnetic dipole in which the flux field is stable. The function defining the rate at which this field-strength decrease occurs depends on the geometry of the magnetic lines of flux (the shape of the flux field).

PREFIXES

Adding a prefix to the name of that unit forms a multiple of a unit in the International System (see the table "International System Prefixes" on page 201). The prefixes change the magnitude of the unit by orders of 10 from 10^{18} to 10^{-18}.

Very small concentrations of chemicals are also measured in parts per million (ppm) or parts per billion (ppb), which mean just what they sound like: If there are four parts per million of lead in a rock (4 ppm), then out of every million atoms in that rock, on average four of them will be lead.

APPENDIX 2:

Light, Wavelength, and Radiation

Electromagnetic radiation is energy given off by matter, traveling in the form of waves or particles. Electromagnetic energy exists in a wide range of energy values, of which visible light is one small part of the total spectrum. The source of radiation may be the hot and therefore highly energized atoms of the Sun, pouring out radiation across a wide range of energy values, including of course visible light, and they may also be unstable (radioactive) elements giving off radiation as they decay.

Radiation is called "electromagnetic" because it moves as interlocked waves of electrical and magnetic fields. A wave is a disturbance traveling through space, transferring energy from one point to the next. In a vacuum, all electromagnetic radiation travels at the speed of light, 983,319,262 feet per second (299,792,458 m/sec, often approximated as 300,000,000 m/sec). Depending on the type of radiation, the waves have different wavelengths, energies, and frequencies (see the following figure). The wavelength is the distance between individual waves, from one peak to another. The frequency is the number of waves that pass a stationary point each second. Notice in the graphic how the wave undulates up and down from peaks to valleys to peaks. The time from one peak to the next peak is called one cycle. A single unit of frequency is equal to one cycle per second. Scientists refer to a single cycle as one hertz, which commemorates 19th-century German physicist Heinrich Hertz, whose discovery of electromagnetic waves led to the development of radio. The frequency of a wave is related to its energy: The higher the frequency of a wave, the higher its energy, though its speed in a vacuum does not change.

The smallest wavelength, highest energy and frequency electromagnetic waves are cosmic rays, then as wavelength increases and energy and frequency decrease, come gamma rays, then X-rays, then ultraviolet light, then visible light (moving from violet through indigo, blue, green, yellow, orange, and red), then infrared (divided into near, meaning near to visible, mid-, and far infrared), then microwaves, and then radio waves, which have the longest wavelengths and the lowest energy and frequency. The electromagnetic spectrum is shown in the accompanying figure and table on page 206.

As a wave travels and vibrates up and down with its characteristic wavelength, it can be imagined as vibrating up and down in a single plane, such as the plane of this sheet of paper in the case of the simple example in the figure here showing polarization. In nature, some waves change their polarization constantly so that their polarization sweeps through all angles, and they are said to be circularly polarized. In ordinary visible light, the waves are vibrating up and down in numerous random planes. Light can be shone through a special filter called a polarizing filter that blocks out all the light except that polarized in a certain direction, and the light that shines out the other side of the filter is then called polarized light.

Each electromagnetic wave has a measurable wavelength and frequency.

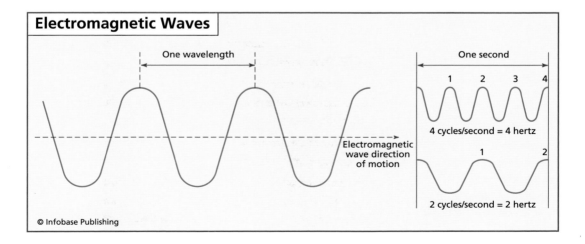

Electromagnetic Waves

One wavelength

One second

1 2 3 4

Electromagnetic wave direction of motion

4 cycles/second = 4 hertz

1 2

2 cycles/second = 2 hertz

Polarization is important in wireless communications systems such as radios, cell phones, and non-cable television. The orientation of the transmitting antenna creates the polarization of the radio waves transmitted by that antenna: A vertical antenna emits vertically polarized waves, and a horizontal antenna emits horizontally polarized waves. Similarly, a horizontal antenna is best at receiving horizontally polarized waves and a vertical antenna at vertically polarized waves.

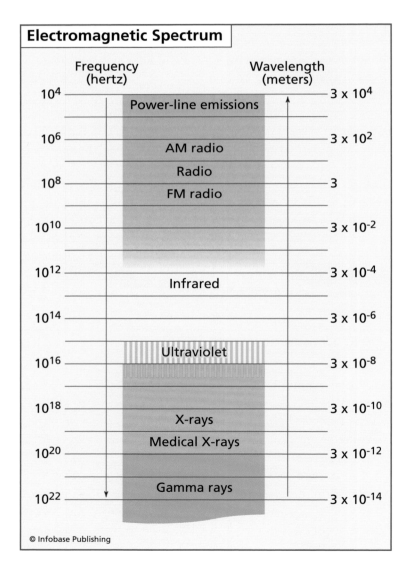

The electromagnetic spectrum ranges from cosmic rays at the shortest wavelengths to radiowaves at the longest wavelengths.

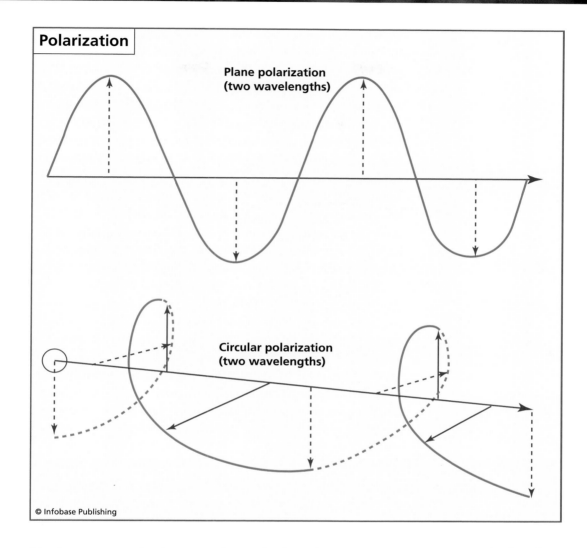

Polarization

Plane polarization
(two wavelengths)

Circular polarization
(two wavelengths)

© Infobase Publishing

The best communications are obtained when the source and receiver antennas have the same polarization. This is why, when trying to adjust television antennas to get a better signal, having the two antennae at right angles to each other can maximize the chances of receiving a signal.

The human eye stops being able to detect radiation at wavelengths between 3,000 and 4,000 angstroms, which is deep violet—also the rough limit on transmissions through the atmosphere (see the table "Wavelengths and Frequencies of Visible Light" on page 208). (Three thousand to 4,000

Waves can be thought of as plane or circularly polarized.

angstroms is the same as 300–400 nm because an angstrom is 10^{-9} m, while the prefix nano- or n means 10^{-10}; for more, see appendix 1, "Units and Measurements.") Of visible light, the colors red, orange, yellow, green, blue, indigo, and violet are listed in order from longest wavelength and lowest energy to shortest wavelength and highest energy. Sir Isaac Newton, the spectacular English physicist and mathematician, first found that a glass prism split sunlight into a rainbow of colors. He named this a "spectrum," after the Latin word for ghost.

If visible light strikes molecules of gas as it passes through the atmosphere, it may get absorbed as energy by the molecule. After a short amount of time, the molecule releases the light, most probably in a different direction. The color that is radiated is the same color that was absorbed. All the colors of visible light can be absorbed by atmospheric molecules, but the higher energy blue light is absorbed more often than the lower energy red light. This process is called Rayleigh scattering (named after Lord John Rayleigh, an English physicist who first described it in the 1870s).

The blue color of the sky is due to Rayleigh scattering. As light moves through the atmosphere, most of the longer wavelengths pass straight through: The air affects little of the red,

WAVELENGTHS AND FREQUENCIES OF VISIBLE LIGHT

Visible light color	Wavelength (in Å, angstroms)	Frequency (times 10^{14} Hz)
violet	4,000–4,600	7.5–6.5
indigo	4,600–4,750	6.5–6.3
blue	4,750–4,900	6.3–6.1
green	4,900–5,650	6.1–5.3
yellow	5,650–5,750	5.3–5.2
orange	5,750–6,000	5.2–5.0
red	6,000–8,000	5.0–3.7

WAVELENGTHS AND FREQUENCIES OF THE ELECTROMAGNETIC SPECTRUM

Energy	Frequency in hertz (Hz)	Wavelength in meters
cosmic rays	everything higher in energy than gamma rays	everything lower in wavelength than gamma rays
gamma rays	10^{20} to 10^{24}	less than 10^{-12} m
X-rays	10^{17} to 10^{20}	1 nm to 1 pm
ultraviolet	10^{15} to 10^{17}	400 nm to 1 nm
visible	4×10^{14} to 7.5×10^{14}	750 nm to 400 nm
near-infrared	1×10^{14} to 4×10^{14}	2.5 μm to 750 nm
infrared	10^{13} to 10^{14}	25 μm to 2.5 μm
microwaves	3×10^{11} to 10^{13}	1 mm to 25 μm
radio waves	less than 3×10^{11}	more than 1 mm

orange, and yellow light. The gas molecules absorb much of the shorter wavelength blue light. The absorbed blue light is then radiated in different directions and is scattered all around the sky. Whichever direction you look, some of this scattered blue light reaches you. Since you see the blue light from everywhere overhead, the sky looks blue. Note also that there is a very different kind of scattering, in which the light is simply bounced off larger objects like pieces of dust and water droplets, rather than being absorbed by a molecule of gas in the atmosphere and then reemitted. This bouncing kind of scattering is responsible for red sunrises and sunsets.

Until the end of the 18th century, people thought that visible light was the only kind of light. The amazing amateur astronomer Frederick William Herschel (the discoverer of Uranus) discovered the first non-visible light, the infrared. He thought that each color of visible light had a different temperature and devised an experiment to measure the temperature of each color of light. The temperatures went up as the colors progressed from violet through red, and then Herschel decided

COMMON USES FOR RADIO WAVES

User	Approximate frequency
AM radio	0.535×10^6 to 1.7×10^6Hz
baby monitors	49×10^6Hz
cordless phones	49×10^6Hz
	900×10^6Hz
	$2,400 \times 10^6$Hz
television channels 2 through 6	54×10^6 to 88×10^6Hz
radio-controlled planes	72×10^6Hz
radio-controlled cars	75×10^6Hz
FM radio	88×10^6 to 108×10^6Hz
television channels 7 through 13	174×10^6 to 220×10^6Hz
wildlife tracking collars	215×10^6Hz
cell phones	800×10^6Hz
	$2,400 \times 10^6$Hz
air traffic control radar	960×10^6Hz
	$1,215 \times 10^6$Hz
global positioning systems	$1,227 \times 10^6$Hz
	$1,575 \times 10^6$Hz
deep space radio	$2,300 \times 10^6$Hz

to measure past red, where he found the highest temperature yet. This was the first demonstration that there was a kind of radiation that could not be seen by the human eye. Herschel originally named this range of radiation "calorific rays," but the name was later changed to infrared, meaning "below red." Infrared radiation has become an important way of sensing solar system objects and is also used in night-vision goggles and various other practical purposes.

At lower energies and longer wavelengths than the visible and infrared, microwaves are commonly used to transmit energy to food in microwave ovens, as well as for some communications, though radio waves are more common in this use. There is a wide range of frequencies in the radio spectrum, and they are used in many ways, as shown in the table "Common Uses for Radio Waves," including television, radio, and cell phone transmissions. Note that the frequency units are given in terms of 10^6 Hz, without correcting for each coefficient's additional factors of 10. This is because 10^6 Hz corresponds to the unit of megahertz (MHz), which is a commonly used unit of frequency.

Cosmic rays, gamma rays, and X-rays, the three highest-energy radiations, are known as ionizing radiation because they contain enough energy that, when they hit an atom, they may knock an electron off of it or otherwise change the atom's weight or structure. These ionizing radiations, then, are particularly dangerous to living things; for example, they can damage DNA molecules (though good use is made of them as well, to see into bodies with X-rays and to kill cancer cells with gamma rays). Luckily the atmosphere stops most ionizing radiation, but not all of it. Cosmic rays created by the Sun in solar flares, or sent off as a part of the solar wind, are relatively low energy. There are far more energetic cosmic rays, though, that come from distant stars through interstellar space. These are energetic enough to penetrate into an asteroid as deeply as a meter and can often make it through the atmosphere.

When an atom of a radioisotope decays, it gives off some of its excess energy as radiation in the form of X-rays, gamma rays, or fast-moving subatomic particles: alpha particles (two protons and two neutrons, bound together as an atomic *nucleus*), or beta particles (fast-moving electrons), or a combination of two or more of these products. If it decays with emission of an alpha or beta particle, it becomes a new element. These decay products can be described as gamma, beta, and alpha radiation. By decaying, the atom is progressing in

RADIOACTIVITY OF SELECTED OBJECTS AND MATERIALS

Object or material	Radioactivity
1 adult human (100 Bq/kg)	7,000 Bq
1 kg coffee	1,000 Bq
1 kg high-phosphate fertilizer	5,000 Bq
1 household smoke detector (with the element americium)	30,000 Bq
radioisotope source for cancer therapy	100 million million Bq
1 kg 50-year-old vitrified high-level nuclear waste	10 million million Bq
1 kg uranium ore (Canadian ore, 15% uranium)	25 million Bq
1 kg uranium ore (Australian ore, 0.3% uranium)	500,000 Bq
1 kg granite	1,000 Bq

one or more steps toward a stable state where it is no longer radioactive.

The X-rays and gamma rays from decaying atoms are identical to those from other natural sources. Like other ionizing radiation, they can damage living tissue but can be blocked by lead sheets or by thick concrete. Alpha particles are much larger and can be blocked more quickly by other material; a sheet of paper or the outer layer of skin on your hand will stop them. If the atom that produces them is taken inside the body, however, such as when a person breathes in radon gas, the alpha particle can do damage to the lungs. Beta particles are more energetic and smaller and can penetrate a couple of centimeters into a person's body.

But why can both radioactive decay that is formed of sub-atomic particles and heat that travels as a wave of energy be considered radiation? One of Albert Einstein's great discoveries is called the photoelectric effect: Subatomic particles can all behave as either a wave or a particle. The smaller the particle, the more wavelike it is. The best example of this is light itself, which behaves almost entirely as a wave, but there is the particle equivalent for light, the massless photon. Even alpha particles, the largest decay product discussed here, can act like a wave, though their wavelike properties are much harder to detect.

The amount of radioactive material is given in becquerel (Bq), a measure that enables us to compare the typical radioactivity of some natural and other materials. A becquerel is one atomic decay per second. Radioactivity is still sometimes measured using a unit called a Curie; a Becquerel is 27×10^{-12} Curies. There are materials made mainly of radioactive elements, like uranium, but most materials are made mainly of stable atoms. Even materials made mainly of stable atoms, however, almost always have trace amounts of radioactive elements in them, and so even common objects give off some level of radiation, as shown in the table on page 212.

Background radiation is all around us all the time. Naturally occurring radioactive elements are more common in some kinds of rocks than others; for example, *granite* carries more radioactive elements than does sandstone; therefore a person working in a bank built of granite will receive more radiation than someone who works in a wooden building. Similarly, the atmosphere absorbs cosmic rays, but the higher the elevation, the more cosmic-ray exposure there is. A person living in Denver or in the mountains of Tibet is exposed to more cosmic rays than someone living in Boston or in the Netherlands.

APPENDIX 3:

A List of All Known Moons

Though Mercury and Venus have no moons, the other planets in the solar system have at least one. Some moons, such as Earth's Moon and Jupiter's Galileans satellites, are thought to have formed at the same time as their accompanying planet. Many other moons appear simply to be captured asteroids; for at least half of Jupiter's moons, this seems to be the case. These small, irregular moons are difficult to detect from Earth, and so the lists given in the table below must be considered works in progress for the gas giant planets. More moons will certainly be discovered with longer observation and better instrumentation.

KNOWN MOONS OF ALL PLANETS						
Earth	Mars	Jupiter	Saturn	Uranus	Neptune	Pluto
1	2	63	62	27	13	3
1. Moon	1. Phobos	1. Metis	1. S/2009 S1	1. Cordelia	1. Naiad	1. Charon
	2. Diemos	2. Adrastea	2. Pan	2. Ophelia	2. Thalassa	2. Nix (P1)
		3. Amalthea	3. Daphnis	3. Bianca	3. Despina	3. Hydra
		4. Thebe	4. Atlas	4. Cressida	4. Galatea	(P2)
		5. Io	5. Prometheus	5. Desdemona	5. Larissa	
		6. Europa	6. Pandora	6. Juliet	6. Proteus	
		7. Ganymede	7. Epimetheus	7. Portia	7. Triton	
		8. Callisto	8. Janus	8. Rosalind	8. Nereid	
		9. Themisto	9. Aegaeon	9. Cupid (2003	9. Halimede	
		10. Leda	10. Mimas	U2)	(S/2002	
		11. Himalia	11. Methone	10. Belinda	N1)	

Earth	Mars	Jupiter	Saturn	Uranus	Neptune	Pluto
		12. Lysithea	12. Anthe	11. Perdita (1986 U10)	10. Sao (S/2002 N2)	
		13. Elara	13. Pallene	12. Puck	11. Laomedeia (S/2002 N3)	
		14. S/2000 J11	14. Enceladus	13. Mab (2003 U1)	12. Psamathe (S/2003 N1)	
		15. Carpo (S/2003 J20)	15. Telesto	14. Miranda	13. Neso (S/2002 N4)	
		16. S/2003 J12	16. Tethys	15. Ariel		
		17. Euporie	17. Calypso	16. Umbriel		
		18. S/2003 J3	18. Dione	17. Titania		
		19. S/2003 J18	19. Helene	18. Oberon		
		20. Orthosie	20. Polydeuces	19. Francisco (2001 U3)		
		21. Euanthe	21. Rhea	20. Caliban		
		22. Harpalyke	22. Titan	21. Stephano		
		23. Praxidike	23. Hyperion	22. Trinculo		
		24. Thyone	24. Iapetus	23. Sycorax		
		25. S/2003 J16	25. Kiviuq	24. Margaret (2003 U3)		
		26. Mneme (S/2003 J21)	26. Ijiraq	25. Prospero		
		27. Iocaste	27. Phoebe	26. Setebos		
		28. Helike (S/2003 J6)	28. Paaliaq	27. Ferdinand (2001 U2)		
		29. Hermippe	29. Skathi			
		30. Thelxinoe (S/2003 J22)	30. Albiorix			
		31. Ananke	31. S/2007 S2			
		32. S/2003 J15	32. Bebhionn			
		33. Eurydome	33. Erriapo			
		34. S/2003 J17	34. Siarnaq			
		35. Pasithee	35. Skoll			
		36. S/2003 J10	36. Tarvos			
			37. Tarqeq			
			38. Greip			
			39. Hyrrokkin			
			40. S/2004 S13			
			41. S/2004 S17			
			42. Mundilfari			
			43. Jarnsaxa			
			44. S/2006 S1			

(continues)

KNOWN MOONS OF ALL PLANETS *(continued)*

Earth	Mars	Jupiter	Saturn	Uranus	Neptune	Pluto
		37. Chaldene	45. Narvi			
		38. Isonoe	46. Bergelmir			
		39. Erinome	47. Suttungr			
		40. Kale	48. S/2004 S12			
		41. Aitne	49. S/2004 S7			
		42. Taygete	50. Hati			
		43. Kallichore (S/2003 J11)	51. Bestla			
		44. Eukelade (S/2003 J1)	52. Farbauti			
		45. Arche (S/2002 J1)	53. Thrymyr			
		46. S/2003 J9	54. S/2007 S3			
		47. Carme	55. Aegir			
		48. Kalyke	56. S/2006 S3			
		49. Sponde	57. Kari			
		50. Magaclite	58. Fenrir			
		51. S/2003 J5	59. Surtur			
		52. S/2003 J19	60. Ymir			
		53. S/2003 J23	61. Loge			
		54. Hegemone (S/2003 J8)	62. Fornjot			
		55. Pasiphae				
		56. Cyllene (S/2003 J13)				
		57. S/2003 J4				
		58. Sinope				
		59. Aoede (S/2003 J7)				
		60. Autonoe				
		61. Calirrhoe				
		62. Kore (S/2003 J14)				
		63. S/2003 J2				

Glossary

accretion The accumulation of celestial gas, dust, or smaller bodies by gravitational attraction into a larger body, such as a planet or an asteroid

achondite A stony (silicate-based) meteorite that contains no chondrules; these originate in differentiated bodies and may be mantle material or lavas (see also chondrite and iron meteorite)

albedo The light reflected by an object as a fraction of the light shining on an object; mirrors have high albedo, while charcoal has low albedo

anorthite A calcium-rich plagioclase mineral with compositional formula $CaAl_2Si_2O_8$, significant for making up the majority of the rock anorthosite in the crust of the Moon

anticyclone An area of increased atmospheric pressure relative to the surrounding pressure field in the atmosphere, resulting in circular flow in a clockwise direction north of the equator and in a counterclockwise direction to the south

aphelion A distance; the farthest from the Sun an object travels in its orbit

apogee As for aphelion but for any orbital system (not confined to the Sun)

apparent magnitude The brightness of a celestial object as it would appear from a given distance—the lower the number, the brighter the object

atom The smallest quantity of an element that can take part in a chemical reaction; consists of a nucleus of protons and neutrons, surrounded by a cloud of electrons; each atom is about 10^{-10} meters in diameter, or one angstrom

atomic number   The number of protons in an atom's nucleus

AU   An AU is an astronomical unit, defined as the distance from the Sun to the Earth; approximately 93 million miles, or 150 million kilometers. For more information, refer to the UNITS AND MEASUREMENTS appendix

basalt   A generally dark-colored extrusive igneous rock most commonly created by melting a planet's mantle; its low silica content indicates that it has not been significantly altered on its passage to the planet's surface

bolide   An object falling into a planet's atmosphere, when a specific identification as a comet or asteroid cannot be made

bow shock   The area of compression in a flowing fluid when it strikes an object or another fluid flowing at another rate; for example, the bow of a boat and the water, or the magnetic field of a planet and the flowing solar wind

breccia   Material that has been shattered from grinding, as in a fault, or from impact, as by meteorites or other solar system bodies

CAIs   Calcium-aluminum inclusions, small spheres of mineral grains found in chondritic meteorites and thought to be the first solids that formed in the protoplanetary disk

calcium-aluminum inclusion   See CAIs

chondrite   A class of meteorite thought to contain the most primitive material left from the solar nebula; named after their glassy, super-primitive inclusions called chondrules

chondrule   Rounded, glassy, and crystalline bodies incorporated into the more primitive of meteorites; thought to be the condensed droplets of the earliest solar system materials

CI chondrite   The class on chondrite meteorites with compositions most like the Sun, and therefore thought to be the oldest and least altered material in the solar system

clinopyroxene   A common mineral in the mantle and igneous rocks, with compositional formula $((Ca,Mg,Fe,Al)_2(Si,Al)_2O_6)$

conjunction When the Sun is between the Earth and the planet or another body in question

convection Material circulation upward and downward in a gravity field caused by horizontal gradients in density; an example is the hot, less dense bubbles that form at the bottom of a pot, rise, and are replaced by cooler, denser sinking material

core The innermost material within a differentiated body; in a rocky planet this consists of iron-nickel metal, and in a gas planet this consists of the rocky innermost solids

Coriolis force The effect of movement on a rotating sphere; movement in the Northern Hemisphere curves to the right, while movement in the Southern Hemisphere curves to the left

craton The ancient, stable interior cores of the Earth's continents

crust The outermost layer of most differentiated bodies, often consisting of the least dense products of volcanic events or other buoyant material

cryovolcanism Non-silicate materials erupted from icy and gassy bodies in the cold outer solar system; for example, as suspected or seen on the moons Enceladus, Europa, Titan, and Triton

cubewano Any large Kuiper belt object orbiting between about 41 AU and 48 AU but not controlled by orbital resonances with Neptune; the odd name is derived from 1992 QB_1, the first Kuiper belt object found

cyclone An area in the atmosphere in which the pressures are lower than those of the surrounding region at the same level, resulting in circular motion in a counterclockwise direction north of the equator and in a clockwise direction to the south

debris disk A flattened, spinning disk of dust and gas around a star formed from collisions among bodies already accreted in an aging solar system

differential rotation   Rotation at different rates at different latitudes, requiring a liquid or gassy body, such as the Sun or Jupiter

differentiated body   A spherical body that has a structure of concentric spherical layers, differing in terms of composition, heat, density, and/or motion; caused by gravitational separations and heating events such as planetary accretion

dipole   Two associated magnetic poles, one positive and one negative, creating a magnetic field

direct (prograde)   Rotation or orbit in the same direction as the Earth's, that is, counterclockwise when viewed from above its North Pole

disk wind   Magnetic fields that either pull material into the protostar or push it into the outer disk; these are thought to form at the inner edge of the disk where the protostar's magnetic field crosses the disk's magnetic field (also called **x-wind**)

distributary   River channels that branch from the main river channel, carrying flow away from the central channel; usually form fans of channels at a river's delta

eccentricity   The amount by which an ellipse differs from a circle

ecliptic   The imaginary plane that contains the Earth's orbit and from which the planes of other planets' orbits deviate slightly (Pluto the most, by 17 degrees); the ecliptic makes an angle of 7 degrees with the plane of the Sun's equator

ejecta   Material thrown out of the site of a crater by the force of the impactor

element   A family of atoms that all have the same number of positively charged particles in their nuclei (the center of the atom)

ellipticity   The amount by which a planet's shape deviates from a sphere

equinox   One of two points in a planet's orbit when day and night have the same length; vernal equinox occurs in Earth's spring and autumnal equinox in the fall

exosphere The uppermost layer of a planet's atmosphere

extrasolar Outside this solar system

faint young Sun paradox The apparent contradiction between the observation that the Sun gave off far less heat in its early years, and the likelihood that the Earth was still warm enough to host liquid water

garnet The red, green, or purple mineral that contains the majority of the aluminum in the Earth's upper mantle; its compositional formula is $((Ca,Mg,Fe Mn)_3(Al,Fe,Cr,Ti)_2(SiO_4)3)$

giant molecular cloud An interstellar cloud of dust and gas that is the birthplace of clusters of new stars as it collapses through its own gravity

graben A low area longer than it is wide and bounded from adjoining higher areas by faults; caused by extension in the crust

granite An intrusive igneous rock with high silica content and some minerals containing water; in this solar system thought to be found only on Earth

half-life The time it takes for half a population of an unstable isotope to decay

hydrogen burning The most basic process of nuclear fusion in the cores of stars that produces helium and radiation from hydrogen

igneous rock Rock that was once hot enough to be completely molten

impactor A generic term for the object striking and creating a crater in another body

inclination As commonly used in planetary science, the angle between the plane of a planet's orbit and the plane of the ecliptic

iron meteorite Meteorites that consist largely of iron-nickel metal; thought to be parts of the cores of smashed planetesimals from early solar system accretion

isotope Atoms with the same number of protons (and are therefore the same type of element) but different numbers of neutrons; may be stable or radioactive and occur in different relative abundances

lander   A spacecraft designed to land on another solar system object rather than flying by, orbiting, or entering the atmosphere and then burning up or crashing

lithosphere   The uppermost layer of a terrestrial planet consisting of stiff material that moves as one unit if there are plate tectonic forces and does not convect internally but transfers heat from the planet's interior through conduction

magnetic moment   The torque (turning force) exerted on a magnet when it is placed in a magnetic field

magnetopause   The surface between the magnetosheath and the magnetosphere of a planet

magnetosheath   The compressed, heated portion of the solar wind where it piles up against a planetary magnetic field

magnetosphere   The volume of a planet's magnetic field, shaped by the internal planetary source of the magnetism and by interactions with the solar wind

magnitude   See APPARENT MAGNITUDE

mantle   The spherical shell of a terrestrial planet between crust and core; thought to consist mainly of silicate minerals

mass number   The number of protons plus neutrons in an atom's nucleus

mesosphere   The atmospheric layer between the stratosphere and the thermosphere

metal   1) Material with high electrical conductivity in which the atomic nuclei are surrounded by a cloud of electrons, that is, metallic bonds, or 2) In astronomy, any element heavier than helium

metallicity   The fraction of all elements heavier than hydrogen and helium in a star or protoplanetary disk; higher metallicity is thought to encourage the formation of planets

metamorphic rock   Rock that has been changed from its original state by heat or pressure but was never liquid

mid-ocean ridge   The line of active volcanism in oceanic basins from which two oceanic plates are produced, one moving away from each side of the ridge; only exist on Earth

mineral   A naturally occurring inorganic substance having an orderly internal structure (usually crystalline) and characteristic chemical composition

nucleus   The center of the atom, consisting of protons (positively charged) and neutrons (no electric charge); tiny in volume but makes up almost all the mass of the atom

nutation   The slow wobble of a planet's rotation axis along a line of longitude, causing changes in the planet's obliquity

obliquity   The angle between a planet's equatorial plane to its orbit plane

occultation   The movement of one celestial body in front of another from a particular point of view; most commonly the movement of a planet in front of a star from the point of view of an Earth viewer

olivine   Also known as the gem peridot, the green mineral that makes up the majority of the upper mantle; its compositional formula is $((Mg, Fe)_2SiO_4)$

one-plate planet   A planet with lithosphere that forms a continuous spherical shell around the whole planet, not breaking into plates or moving with tectonics; Mercury, Venus, and Mars are examples

opposition   When the Earth is between the Sun and the planet of interest

orbital period   The time required for an object to make a complete circuit along its orbit

pallasite   A type of iron meteorite that also contains the silicate mineral olivine, and is thought to be part of the region between the mantle and core in a differentiated planetesimal that was shattered in the early years of the solar system

parent body   The larger body that has been broken to produce smaller pieces; large bodies in the asteroid belt are thought to be the parent bodies of meteorites that fall to Earth today

perigee As for perihelion but for any orbital system (not confined to the Sun)

perihelion A distance; the closest approach to the Sun made in an object's orbit

planetary nebula A shell of gas ejected from stars at the end of their lifetimes; unfortunately named in an era of primitive telescopes that could not discern the size and nature of these objects

planetesimal The small, condensed bodies that formed early in the solar system and presumably accreted to make the planets; probably resembled comets or asteroids

plate tectonics The movement of lithospheric plates relative to each other, only known on Earth

precession The movement of a planet's axis of rotation that causes the axis to change its direction of tilt, much as the direction of the axis of a toy top rotates as it slows

primordial disk Another name for a protoplanetary disk

prograde (direct) Rotates or orbits in the same direction the Earth does, that is, counterclockwise when viewed from above its North Pole

proplyd Abbreviation for a *protoplanetary disk*

protoplanetary disk The flattened, spinning cloud of dust and gas surrounding a growing new star

protostar The central mass of gas and dust in a newly forming solar system that will eventually begin thermonuclear fusion and become a star

radioactive An atom prone to radiodecay

radio-decay The conversion of an atom into a different atom or isotope through emission of energy or subatomic particles

red, reddened A solar system body with a redder color in visible light, but more important, one that has increased albedo at low wavelengths (the "red" end of the spectrum)

reflectance spectra   The spectrum of radiation that bounces off a surface, for example, sunlight bouncing off the surface of as asteroid; the wavelengths with low intensities show the kinds of radiation absorbed rather than reflected by the surface, and indicate the composition of the surface materials

refractory   An element that requires unusually high temperatures in order to melt or evaporate; compare to volatile

relief (topographic relief)   The shapes of the surface of land; most especially the high parts such as hills or mountains

resonance   When the ratio of the orbital periods of two bodies is an integer; for example, if one moon orbits its planet once for every two times another moon orbits, the two are said to be in resonance

retrograde   Rotates or orbits in the opposite direction to Earth, that is, clockwise when viewed from above its North Pole

Roche limit   The radius around a given planet that a given satellite must be outside of in order to remain intact; within the Roche limit, the satellite's self-gravity will be overcome by gravitational tidal forces from the planet, and the satellite will be torn apart

rock   Material consisting of the aggregate of minerals

sedimentary rock   Rock made of mineral grains that were transported by water or air

seismic waves   Waves of energy propagating through a planet, caused by earthquakes or other impulsive forces, such as meteorite impacts and human-made explosions

semimajor axis   Half the widest diameter of an orbit

semiminor axis   Half the narrowest diameter of an orbit

silicate   A molecule, crystal, or compound made from the basic building block silica (SiO_2); the Earth's mantle is made of silicates, while its core is made of metals

spectrometer   An instrument that separates electromagnetic radiation, such as light, into wavelengths, creating a spectrum

stratosphere The layer of the atmosphere located between the troposphere and the mesosphere, characterized by a slight temperature increase and absence of clouds

subduction Movement of one lithospheric plate beneath another

subduction zone A compressive boundary between two lithospheric plates, where one plate (usually an oceanic plate) is sliding beneath the other and plunging at an angle into the mantle

synchronous orbit radius The orbital radius at which the satellite's orbital period is equal to the rotational period of the planet; contrast with synchronous rotation

synchronous rotation When the same face of a moon is always toward its planet, caused by the period of the moon's rotation about its axis being the same as the period of the moon's orbit around its planet; most moons rotate synchronously due to tidal locking

tacholine The region in the Sun where differential rotation gives way to solid-body rotation, creating a shear zone and perhaps the body's magnetic field as well; is at the depth of about one-third of the Sun's radius

terrestrial planet A planet similar to the Earth—rocky and metallic and in the inner solar system; includes Mercury, Venus, Earth, and Mars

thermosphere The atmospheric layer between the mesosphere and the exosphere

tidal locking The tidal (gravitational) pull between two closely orbiting bodies that causes the bodies to settle into stable orbits with the same faces toward each other at all times; this final stable state is called synchronous rotation

tomography The technique of creating images of the interior of the Earth using the slightly different speeds of earthquake waves that have traveled along different paths through the Earth

tropopause The point in the atmosphere of any planet where the temperature reaches a minimum; both above and below this height, temperatures rise

troposphere   The lower regions of a planetary atmosphere, where convection keeps the gas mixed, and there is a steady decrease in temperature with height above the surface

viscosity   A liquid's resistance to flowing; honey has higher viscosity than water

visual magnitude   The brightness of a celestial body as seen from Earth categorized on a numerical scale; the brightest star has magnitude −1.4 and the faintest visible star has magnitude 6; a decrease of one unit represents an increase in brightness by a factor of 2.512; system begun by Ptolemy in the second century B.C.E.; see also apparent magnitude

volatile   An element that moves into a liquid or gas state at relatively low temperatures; compare with refractory

x-wind   Magnetic fields that either pull material into the protostar or push it into the outer disk; these are thought to form at the inner edge of the disk where the protostar's magnetic field crosses the disk's magnetic field (also called **disk wind**)

Further Resources

Arnett, Bill. "Discovery and Origins of Pluto." *Lunar and Planetary Information Bulletin* 9: 1–7 (2001). Simple, brief summary of the state of knowledge about Pluto in 2001.

Beatty, J. K., C. C. Petersen, and A. Chaikin. *The New Solar System*. Cambridge: Sky Publishing and Cambridge University Press, 1999. The best-known and best-regarded single reference volume on the solar system.

Bissell, Tom. "A Comet's Tale." *Harper's Magazine,* February 2003. Accessible article on comets.

Booth, N. *Exploring the Solar System*. Cambridge: Cambridge University Press, 1995. Well-written and accurate volume on solar system exploration.

Brown, Michael E. "Pluto and Charon: Formation, Seasons, Composition." *Annual Reviews of the Earth and Planetary Sciences* 30 (2002): 307–345. Complete summary of scientific knowledge of Pluto.

———— "The Kuiper Belt." *Physics Today* (April 2004): 49–54. Overview of the Kuiper belt for the lay scientific audience.

Comins, Neil F. and William J. Kaufmann. *Discovering the Universe*. New York: W. H. Freeman, 2008. The best-selling text for astronomy courses that use no mathematics. Presents concepts clearly and stresses the process of science.

Dickin, A.P. *Radiogenic Isotope Geology*. Cambridge: Cambridge University Press, 1995. Thorough college-level text on the uses of radiogenic isotope systems in geology and planetary science.

Elliot, J. L., E. W. Dunham, and D. J. Mink. "The Rings of Uranus." *Nature* 267 (1977): 328–330. Discovery article for Uranus's rings.

Fradin, Dennis Brindell. *The Planet Hunters: The Search for Other Worlds*. New York: Simon & Schuster, 1997. Stories of the people who through time have hunted for and found the planets.

Herschel, W. "Observations on the Two Lately Discovered Celestial Bodies." *Philosophical Transactions of the Royal Society.* London 2 (1802). An original paper by the discoverer of Uranus and several moons and asteroids.

Hofstadter, M. D., and B. J. Butler. "Seasonal Changes in the Deep Atmosphere of Uranus." *Icarus* 165 (2003): 168–180. Scientific article on discoveries in Uranus's deep atmosphere.

Hubbard, W. B., A. Brahic, B. Sicardy, L. R. Elicer, F. Roques, and F. Vilas. "Occultation Detection of a Neptunian Ring-like Arc." *Nature* 319 (1986): 636–640. Discovery article for Neptune's rings.

Luu, Jane X. and David C. Jewitt. "Kuiper Belt Objects: Relics from the Accretion Disk of the Sun." *Annual Reviews of the Earth and Planetary Sciences* 40 (2002): 63–101. Thorough review of the state of knowledge of the Kuiper belt.

Norton, O. R. *The Cambridge Encyclopedia of Meteorites.* Cambridge: Cambridge University Press, 2002. Complete and well-illustrated encyclopedia of meteorites, scientifically correct and yet accessible.

Paul, N. *The Solar System.* Chartwell Books, 2008. Begins with the origin of the universe and moves through the planets. Includes history of space flight and many color images.

Rees, Martin, ed. *Universe.* London: DK Adult, 2005. A team of science writers and astronomers wrote this text for high school students and the general public.

Rivkin, A. *Asteroids, Comets, and Dwarf Planets.* Portsmouth: Heinemann Educational Books, 2009. Up-to-date understanding of the small bodies of the solar system written by an active academic expert in the field.

Schmadel, L. D. *Dictionary of Minor Planet Names.* Berlin: Springer-Verlag, 1999. Complete list as of its publication date, including explanations for naming conventions.

Sparrow, Giles. *The Planets: A Journey Through the Solar System.* Waltham, Mass.: Quercus Press, 2009. Solar system discoveries told within the structure of the last 40 years of space missions.

Spence, P. *The Universe Revealed.* Cambridge: Cambridge University Press, 1998. Comprehensive textbook on the universe.

Stacey, Frank D. *Physics of the Earth.* Brisbane, Australia: Brookfield Press, 1992. Fundamental geophysics text on an upper-level undergraduate college level.

Stern, Alan, and Hal Levison. "Toward a Planet Paradigm." *Sky and Telescope* (August 2002): 42–46. Opinions from top scientists in the field about how planets should be categorized.

Stevenson, D. J. "Planetary Magnetic Fields." *Earth and Planetary Science Letters,* 208 (2003): 1–11. Comparisons and calculations about the planetary magnetic fields of many bodies in our solar system.

Thommes, E. W., M. J. Duncan, and H. F. Levison. "The Formation of Uranus and Neptune Among Jupiter and Saturn." *Astronomical Journal* 123 (2002): 2,862–2,883. Scientific journal article about how these smaller planets might have managed to form so near the larger, dominating planets.

Trujillo, Chadwick A. "Discovering the Edge of the Solar System." *American Scientist* 91 (2003): 424–431. Accessible article about how the scientists found the first Kuiper belt objects.

Wetherill, G. W. "Provenance of the Terrestrial Planets." *Geochimica et Cosmochimica Acta* 58 (1994): 4,513–4,520. Groundbreaking scientific article on how rocky planets like the Earth are formed.

INTERNET RESOURCES

Arnett, Bill. "The 8 Planets: A Multimedia Tour of the Solar System." Available online. URL: http://nineplanets.org. Accessed September 21, 2009. An accessible overview of the history and science of the nine planets and their moons.

Beisser, K. "*New Horizons:* NASA's Pluto-Kuiper Belt Explorer." Johns Hopkins University Applied Physics Laboratory. Available online. URL: http://pluto.jhuapl.edu/. Accessed November 14, 2009. Details about the mission, images, and educational materials.

Blue, Jennifer, and the Working Group for Planetary System Nomenclature. "Gazetteer of Planetary Nomenclature." United States Geological Survey. Available online. URL: http://planetarynames.wr.usgs.gov/. Accessed September 21, 2009. Complete and official rules for naming planetary features, along with list of all named planetary features and downloadable images.

Boone, Katy, and Brian Dunbar. "*New Horizons* Mission." NASA. Available online. URL: http://www.nasa.gov/mission_pages/newhorizons/main/index.html. Accessed November 14, 2009.

Official NASA site for the mission, with details and news and elapsed mission time and time-to-Pluto-closest-approach clocks.

Brown, Mike. "Mike Brown, Professor of Planetary Astronomy." California Institute of Technology. Available online. URL: http://web.gps.caltech.edu/~mbrown/. Accessed September 21, 2009. Brown's weekly column on planets, details on the dwarf planets including Eris, Sedna, and Quaoar, and details of his research and lectures.

Brown, Michael E. "Sedna (2003 VB12)." California Institute of Technology. Available online. URL: http://www.gps.caltech.edu/~mbrown/sedna/. Accessed September 21, 2009. Narrative Web page about Sedna, possibly the first sighted object in the Oort cloud, prepared by its discoverer.

Chamberlin, Alan, Don Yeomans, Jon Giorgini, Mike Keesey, and Paul Chodas. "Natural Satellite Physical Parameters." Jet Propulsion Laboratory Solar System Dynamics Web site. Available online. URL: http://ssd.jpl.nasa.gov/?sat_phys_par. Accessed September 21, 2009. Physical parameters for all known satellites of all planets.

Jewitt, David. "Kuiper Belt." University of California, Los Angeles. Available online. URL: http://www2.ess.ucla.edu/~jewitt/kb.html. Accessed September 21, 2009. Source of information on Kuiper belt bodies and comets compiled by one of the experts in the field.

Johnston, William Robert. "List of Known Trans-Neptunian Objects." Johnston Archive. Available online. URL: www.johnstonsarchive.net/astro/tnoslist.html. Accessed September 21, 2009. Table of orbital parameters for all known bodies orbiting the Sun beyond Neptune; part of an eclectic scholarly archive on many topics.

LaVoie, Sue, Myche McAuley, and Elizabeth Duxbury Rye. "Planetary Photojournal." Jet Propulsion Laboratory and NASA. Available online. URL: http://photojournal.jpl.nasa.gov/index.html. Accessed September 21, 2009. Large database of public-domain images from space missions.

Lunar and Planetary Institute. "Lunar and Planetary Institute." Universities Space Research Association and NASA. Available online. URL: http://www.lpi.usra.edu/. Accessed September 21, 2009. Wide variety of educational resources on planetary science.

O'Connor, John J., and Edmund F. Robertson. "The MacTutor History of Mathematics Archive." University of St. Andrews, Scotland. Available online. URL: http://www-gap.dcs.st-and. ac.uk/~history/index.html. Accessed September 21, 2009. A scholarly, precise, and eminently accessible compilation of biographies and accomplishments of mathematicians and scientists through the ages.

Rowlett, Russ. "How Many? A Dictionary of Units of Measurement." University of North Carolina at Chapel Hill. Available online. URL: http://www.unc.edu/~rowlett/units. Accessed September 21, 2009. A comprehensive dictionary of units of measurement, from the metric and English systems to the most obscure usages.

Sheppard, Scott S. "Uranus's Known Satellites." Department of Terrestrial Magnetism, Carnegie Institute of Washington. Available online. URL: http://www.dtm.ciw.edu/users/sheppard/satellites/ urasatdata.html. Accessed September 21, 2009. Complete and up-to-date orbital parameters for all known satellites of Uranus.

White, Maura, and Allan Stilwell. "JSC Digital Image Collection." Johnson Space Center. Available online. URL: images.jsc.nasa. gov/index.html. Accessed September 21, 2009. A catalogue of more than 9,000 NASA press release photos from the entirety of the manned space flight program.

Williams, David. "Planetary Fact Sheets." NASA. Available online. URL: http://nssdc.gsfc.nasa.gov/planetary/planetfact.html. Accessed September 21, 2009. Detailed measurements and data on the planets, asteroids, and comets in simple tables.

Williams, David, and Dr. Ed Grayzeck. "Lunar and Planetary Science." NASA. Available online. URL: http://nssdc.gsfc.nasa.gov/ planetary/planetary_home.html. Accessed September 21, 2009. NASA's deep archive and general distribution center for lunar and planetary data and images.

ORGANIZATIONS OF INTEREST

American Geophysical Union (AGU)
2000 Florida Avenue NW
Washington, DC 20009-1277
www.agu.org.
AGU is a worldwide scientific community that advances, through unselfish cooperation in research, the understanding of Earth and space

for the benefit of humanity. AGU is an individual membership society open to those professionally engaged in or associated with the Earth and space sciences. Membership has increased steadily each year, doubling during the 1980s. Membership currently exceeds 41,000, of which about 20 percent are students. Membership in AGU entitles members and associates to receive Eos, *AGU's weekly newspaper, and* Physics Today, *a magazine produced by the American Institute of Physics. In addition, they are entitled to special member rates for AGU publications and meetings.*

Association of Space Explorers
1150 Gemini Avenue
Houston TX 77058
http://www.space-explorers.org/
This association is expressly for people who have flown in space. They include 320 individuals from 34 nations, and their goal is to support space science and education. Their outreach activities include a speakers program, astronaut school visits, and observer status with the United Nations.

European Space Agency (ESA)
8-10 rue Mario Nikis,
75738 Paris, Cedex 15, France
http://www.esa.int/esaCP/index.html
The European Space Agency has 18 member states, and together they create a unified European space program and carry out missions in parallel and in cooperation with NASA, JAXA, and other space agencies. Its member countries are Austria, Belgium, Czech Republic, Denmark, Finland, France, Germany, Greece, Ireland, Italy, Luxembourg, the Netherlands, Norway, Portugal, Spain, Sweden, Switzerland, and the United Kingdom. Hungary, Romania, Poland, and Slovenia are cooperating partners.

International Astronomical Union (IAU)
98bis, bd Arago,
FR-75014
Paris, France,
www.iau.org.
The International Astronomical Union (IAU) was founded in 1919. Its mission is to promote and safeguard the science of astronomy in all its aspects through international cooperation. Its individual members are professional astronomers all over the World, at the Ph.D. level or

beyond and active in professional research and education in astronomy. However, the IAU maintains friendly relations also with organizations that include amateur astronomers in their membership. National Members are generally those with a significant level of professional astronomy. With currently more than 9,100 individual members and 65 national members worldwide, the IAU plays a pivotal role in promoting and coordinating worldwide cooperation in astronomy. The IAU also serves as the internationally recognized authority for assigning designations to celestial bodies and any surface features on them.

Jet Propulsion Laboratory (JPL)

4800 Oak Grove Drive,
Pasadena, California, 91109
www.jpl.nasa.gov.
The Jet Propulsion Laboratory is managed by the California Institute of Technology for NASA. JPL manages many of NASA's space missions, including the Mars Rovers and Cassini, and also conducts fundamental research in planetary and space science.

The Meteoritical Society: The International Society for Meteoritics and Planetary Science

www.meteoriticalsociety.org.
The Meteoritical Society is a nonprofit scholarly organization founded in 1933 to promote the study of extraterrestrial materials and their history. The membership of the society includes 950 scientists and amateur enthusiasts from more than 33 countries who are interested in a wide range of planetary science. Member's interests include meteorites, cosmic dust, asteroids and comets, natural satellites, planets, impacts, and the origins of the solar system.

National Aeronautics and Space Administration (NASA)

300 E Street SW
Washington DC 20002
www.nasa.gov.
NASA, an agency of the United States government, manages space flight centers, research centers, and other organizations, including the National Aerospace Museum. NASA scientists and engineers conduct basic research on planetary and space topics, plan and execute space missions, oversee Earth satellites and data collection, and many other space- and flight-related projects.

The Planetary Society
65 North Catalina Avenue
Pasadena CA 91106-2301
http://www.planetary.org/home/
A Society of lay individuals, scientists, organizations, and businesses dedicated to involving the world's public in space exploration through advocacy, projects, and exploration. The Planetary Society was founded in 1980 by Carl Sagan, Bruce Murray, and Louis Friedman They are particularly dedicated to searching for life outside of the Earth.

Index